LANGE

USMLE ROAD MAP

PHYSIOLOGY

PHYSIOLOGY

Second Edition

JAMES N. PASLEY, PhD

Professor, Physiology and Biophysics and Medical Humanities
Assistant Dean for Educational Advancement
College of Medicine
University of Arkansas for Medical Sciences
Little Rock, Arkansas

Lange Medical Books/McGraw-Hill
Medical Publishing Division

New York Chicago San Francisco Lisbon London Madrid Mexico City
Milan New Delhi San Juan Seoul Singapore Sydney Toronto

The McGraw-Hill Companies

USMLE Road Map: Physiology, Second Edition

1234567890 DOC/DOC 098765

ISBN: 0-07-144517-X

ISSN: 1543-5822

This book was set in Adobe Garamond by Pine Tree Composition, Inc.
The editors were Jason Malley, Christie Naglieri, and Mary E. Bele.
The production supervisor was Sherri Souffrance.
Project management was provided by Pine Tree Composition, Inc.
The index was prepared by Walsh Associates.
RR Donnelley was printer and binder.

This book is printed on acid-free paper

CONTENTS

USING THE
USMLE ROAD MAP SERIES
FOR SUCCESSFUL REVIEW

What Is the Road Map Series?

Short of having your own personal tutor, the USMLE Road Map Series is the best source for efficient review of major concepts and information in the medical sciences.

Why Do You Need A Road Map?

It allows you to navigate quickly and easily through your physiology course notes and textbook and prepares you for USMLE and course examinations.

How Does the Road Map Series Work?

Outline Form: Connects the facts in a conceptual framework so that you understand the ideas and retain the information.

Color and Boldface: Highlights words and phrases that trigger quick retrieval of concepts and facts.

Clear Explanations: Are fine-tuned by years of student interaction. The material is written by authors selected for their excellence in teaching and their experience in preparing students for board examinations.

Illustrations: Provide the vivid impressions that facilitate comprehension and recall.

Clinical Correlations: Link all topics to their clinical applications, promoting fuller understanding and memory retention.

Clinical Problems: Give you valuable practice for the clinical vignette-based USMLE questions.

Explanations of Answers: Are learning tools that allow you to pinpoint your strengths and weaknesses.

For Ruth, Jamie, and Jonathan

Acknowledgments

Special thanks for hard work, technical assistance, good advice, and encouragement to Michael Jennings, Stacy Major, Michael Soulsby, and Richard Wheeler.
This book is dedicated to the many students who have taken medical physiology and USMLE review programs at the University of Arkansas for Medical Sciences College of Medicine.

CHAPTER 1
CELL PHYSIOLOGY

I. Plasma Membrane

A. The surface of the cell is defined by a **plasma membrane** that creates distinct molecular environments within cells. The **lipid bilayer** is similar to thin layers of oil surrounding fluid ozone. Thus, the lipid bilayer divides the cell into functional compartments.

B. The **fluid mosaic model** is the accepted view of the molecular nature of plasma membranes.

1. The model proposes that **proteins traverse the lipid bilayer** and are **incorporated within the lipids.**
2. **Proteins** and **lipids** can move freely in the plane of the membrane, producing the fluid nature of the membrane.

C. The plasma membrane is **composed of phospholipids and proteins.**

1. Membrane lipids can be classified into three major classes: **phospholipids, sphingolipids,** and **cholesterol.**
   a. **Phospholipids** are the most abundant membrane lipids, and phospholipid-bilayer membranes are impermeable to charged molecules.
      (1) They have a **bipolar (amphipathic) nature,** containing a charged head group and two hydrophobic (water-insoluble, noncharged) tails.
      (2) The hydrophobic tails face each other, forming a bilayer and exposing the polar head group to the aqueous environment on either side of the membrane.
   b. **Sphingolipids** have an amphipathic structure similar to phospholipids that allows them to insert into membranes. These lipids can be modified by the addition of carbohydrate units at their polar end, creating **glycosphingolipids** in brain cells.
   c. **Cholesterol** is the predominant sterol (unsaturated alcohols found in animal and plant tissues) in human cells; it increases the fluidity of the membrane by inserting itself between phospholipids, improving membrane stability.

TAY-SACHS DISEASE

The accumulation of glycosphingolipid associated with Tay-Sachs disease causes paralysis and impairment of mental function.

2. Membrane proteins that span the lipid bilayer are known as **integral membrane proteins,** whereas those associated with either the inner or the outer

surface of the plasma membrane are known, respectively, as **peripheral** or **lipid-anchored membrane proteins.**

a. The majority of **integral membrane proteins span the bilayer through the formation of hydrophobic α-helices,** a group of 20–25 amino acids twisted to expose the hydrophobic portion of the amino acids to the lipid environment in the membrane (Figure 1–1).

b. **Protein content** of membranes **varies from less than 20% for myelin,** a substance that helps the propagation of action potentials, to **more than 60% in liver cells,** which perform metabolic activities.

c. **Integral-membrane proteins can act as receptor sites** for antibodies as well as hormone-, neurotransmitter-, and drug-binding sites.

d. **Integral membrane proteins can also be** enzymes that are involved in phosphorylation of metabolic intermediates.

e. Integral membrane proteins can participate in intracellular signaling and growth-regulation pathways.

f. Integral-membrane proteins are involved in the transport movement of water-soluble substances.

g. Integral-membrane proteins can serve as adhesion molecules.

3. **Carrier proteins** in the membrane transport materials across the cell membrane.

4. **Membrane channels** allow polar charged ions (Na^+, K^+, Cl^-, and Ca^{2+}) to flow across the plasma membrane. **Ion channel gates** regulate ion passage and are controlled by voltage (**voltage gated**), ligands (**ligand gated**), or mechanical means (**mechanically gated**).

D. The plasma membrane acts as a selective barrier to maintain the composition of the intracellular environment.

1. **Passive transport,** or diffusion, involves non-coupled transport of solutes across the plasma membrane due to its concentration difference.

a. The term *passive* implies that no energy is expended directly to mediate the transport process.

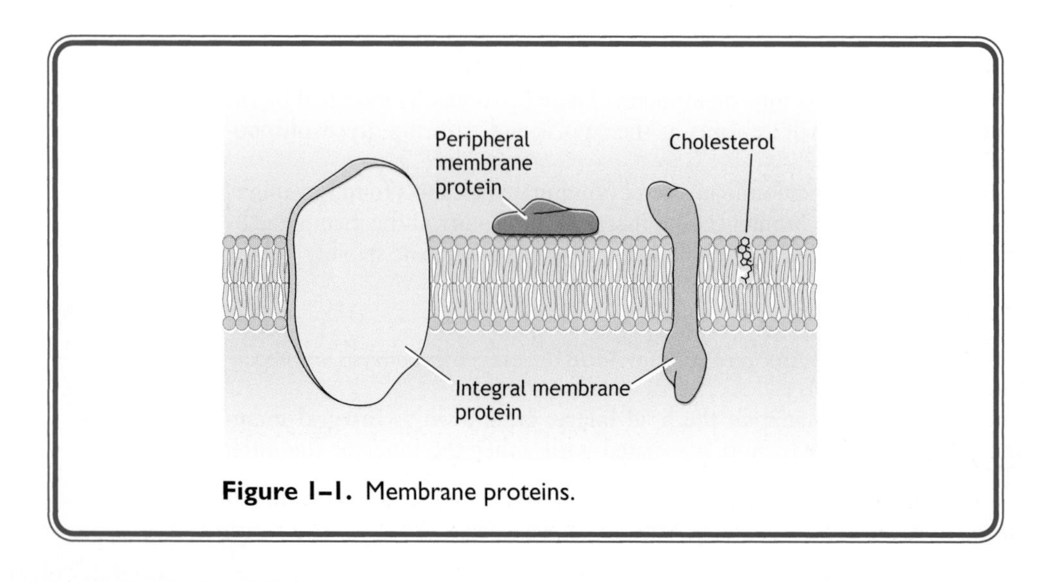

Figure 1–1. Membrane proteins.

b. Passive transport is simple **diffusion** of substances that can readily penetrate the plasma membrane, as is the case for O_2 or CO_2.
c. Passive transport is the only transport mechanism that is not carrier mediated.
d. Substances diffuse because of their inherent random molecular movement (ie, following the principle of **Brownian motion**).
e. Diffusion across membranes occurs if the membrane is permeable to the solute.
f. The net rate of diffusion (J) is proportional to the membrane area (A) and solute concentration difference (C_1–C_2) and the permeability (P) of the membrane.
g. Diffusion is measured using the formula $J = PA(C_1 - C_2)$.

2. **Facilitated diffusion** is the transport of a substrate by a **carrier protein** down its concentration gradient.
   a. Facilitated diffusion is required for substrates that are not permeable to the lipid bilayer and is faster than simple diffusion.
   b. Facilitated diffusion is used to transport a variety of substances required for cellular survival, including glucose and amino acids.
3. **Osmosis** is the movement of water across a semipermeable membrane due to a water concentration difference. Osmosis follows the same principles as the diffusion of any solute.
   a. For example, if two solutions, A and B, are separated by a membrane impermeable to solute but permeable to water and A contains a higher solute concentration than B, a driving force exists for water movement from B to A to equilibrate water concentration differences. Thus, water moves toward a solution with a higher osmolality.
   b. **Osmolality** is a measure of the total concentration of discrete solute particles in solution and is measured in osmoles per kilogram of water.
   c. Because it is much more practical to measure the volume than the weight of physiological solution, the **concentration of solute particles** is typically **expressed as osmolarity,** which is defined as osmoles per liter:

$$\text{Osmolarity} = g \times C$$

   where
   g = number of particles in solution (Osm/mol)
   C = concentration (mol/L)
   d. Consider the following example: What is the osmolarity of a 0.1 mol/L NaCl solution (for NaCl, $g = 2$)?

$$\text{Osmolarity} = 2 \text{ Osm/mol} \times 0.1 \text{ mol/L} = 0.2 \text{ Osm/L or } 200 \text{ mOsm/L}$$

   e. Two solutions that have the same osmolarity are described as **isosmotic.**
4. An **isotonic solution** is one in which the volume of cells incubated in it does not change, implying that there is **no movement of water** in or out of the cell.
   a. Under normal conditions, an isotonic solution is isosmotic with **intracellular fluid,** which is isosmotic with plasma (290 mOsm/L).
   b. Not all isosmotic solutions are isotonic. A 290 mM (millimolar) solution of urea will be isosmotic (290 mOsm/L) but not isotonic because urea is per-

meable to the cell membrane and will diffuse inside the cell. This causes an increased concentration of urea inside the cell, which induces water influx and an increase in cell volume.

5. **Primary active transport** is the transport of a substrate across the plasma membrane against its concentration gradient. It **requires the input of cellular energy** in the form of ATP.
   a. Proteins that mediate primary active transport are known as pumps, which use the energy derived from ATP hydrolysis to power the transport of substrates against their concentration gradient.
   b. The most important example of primary active transport is the **Na^+/K^+-ATPase** or a Na^+/K^+ pump. The Na^+/K^+–ATPase uses the energy of ATP to extrude Na^+ and take up K^+.
   c. Another example of primary active transport is **Ca^{2+}-ATPase,** which clears Ca^{2+} from the cytoplasm. Such Ca^{2+} pumps are found on both the plasma and endoplasmic reticulum (ER) membranes.
   d. In the parietal cells of the gastric gland, an **H-K pump** mediates the active extrusion of H^+ across the apical membrane and the uptake of K^+.
6. **Coupled transport,** or **secondary active transport, uses the energy of ionic gradients,** usually the inwardly directed Na^+ gradient, across the plasma membrane.
   a. Coupled transport still carries substrates against their concentration gradient, but transport is provided indirectly from the energy stored in the concentration gradient of an additional ion transported in the same cycle.
   b. For example, in a Na^+-coupled transporter system, the Na^+ concentration is higher in the extracellular space than in the cytoplasm. Therefore, Na^2 movement into the cytosol is energetically favored.
   c. Coupled transport systems are divided into two groups: **Cotransporters** (also called **symporters**) move solutes in the **same direction,** and **exchangers** (also called **antiporters**) transport solutes in **opposite directions.** Cotransporters and exchangers work only if both substrates are present.
   d. An example of a cotransporter is the **Na^+-glucose transporter,** found in the renal proximal tubule and small intestine, which allows glucose absorption.
   e. An example of an exchanger is the **Na^+-Ca^{2+} exchanger** found in many cell types and important in regulating cytoplasmic Ca^{2+}. The exchanger transports three Na^+ in for one Ca^{2+} out, making it an **electrogenic transporter.** It is electrogenic because it makes a small contribution to the electrical potential across the membrane.

CARDIAC STIMULANTS

- *The Na^+ pump is the target for a class of naturally occurring compounds from the wildflower Digitalis purpurea (foxglove). These compounds have been used for almost two centuries as cardiac stimulants.*
- *These cardiac glycosides, including Ouabain and digitalis, inhibit the Na^+/K^+-ATPase pump.*

II. Ion Channels

A. Ions move quickly through protein pores in biologic membranes known as **ion channels.**

B. Ions flow through these channels from one side of the membrane to the other, down their electrochemical gradients.

C. **Channel proteins** display two different conformational states: open or closed, and are known as gated channels.

D. Gated channels allow ions to cross the membrane passively.

E. **Ion channel gating** is the mechanism that controls the probability of a channel being in each of its conformational states.

1. **Voltage-activated channels** are opened and closed by the membrane potential. For example, a voltage-gated Na^+ channel is closed at the resting membrane potential and is open only when the membrane potential is rapidly depolarized.
2. **Ligand-activated channels** are controlled primarily by the binding of extracellular or intracellular ligands to the channel proteins. These channels are grouped into three categories:
   a. In a **direct receptor channel complex,** the receptor for the ligand is a direct part of the channel protein. The **nicotinic acetylcholine receptor** (AchR) is an example of this type of channel.
   b. In an **intracellular second messenger–gated channel,** the binding of ligands to receptors activates a cascade of second messenger molecules, one of which binds to the channel protein in order to control channel gating. The **cyclic guanosine monophosphate** (cGMP)–gated channel in a photoreceptor is an example.
   c. In a **direct G-protein-gated channel,** the binding of a ligand to its receptor activates a **guanosine triphosphate (GTP)-binding regulatory protein (G-protein)** that changes the conformation of the channel without involving second messenger systems. For example, the cardiac inwardly directed potassium channel K_{Ach}, which slows the heart after vagus nerve stimulation, is gated by a G-protein.

F. Ion channels can select one kind of ion over another.

1. Channels are often named according to the ions they prefer (eg, Na^+ channel, K^+ channel, and Ca^{2+} channels).
2. To account for the selectivity in certain voltage-gated channels, there appears to be a narrow region in the channel pore that fits only on a particular ion.

G. Ion channels provide a useful target for drug action.

1. **Lidocaine,** an antiarrhythmic drug, blocks Na^+ channels in a use-dependent manner.
2. The higher the frequency of stimulation (ie, heart rate), the more that lidocaine blocks the channel.

H. Ion channels are affected by disease both directly and indirectly.

1. **Direct actions** on the channel protein structure occur as a result of genetic mutations of the channel gene.
2. **Indirect actions** include abnormalities in the regulator mechanism required for channel function and in the development of autoimmune disease.

ION CHANNEL DISEASE

- ***Cystic fibrosis** is an autosomal recessive disease that affects 1 in 2500 individuals. It is an example of a **direct** effect on ion channels.*
  *–The disease is caused by mutations in the **cystic fibrosis transmembrane regulator** (CFTR) gene, which codes for the chloride channel gated by **cyclic adenosine monophosphate** (cAMP).*
  –In most cases the deletion of a single phenylalanine molecule (pheD508) prevents the channel protein from reaching the plasma membrane.

–The drastic reduction in chloride channels results in thick mucous secretions that block airways, leading to death in 90% of patients before they reach adulthood.

I. **Cell volume regulation** depends on the total amount of intracellular solute.
   1. Following cell shrinkage, mechanisms that increase solute concentration are activated.
      a. This activation is achieved either by the **synthesis of small organic** (ie, osmotically active) molecules (eg, sorbitol or taurine) or by the **transport of ions** inside the cell through the Na^+-H^+ exchanger or the Na^+-H^+-Cl^- cotransporter.
      b. Increased solute concentration inside the cell will induce water movement by osmosis, increasing cell volume.
      c. Because of the presence of impermeable negatively charged proteins within the cell, osmotic water movement will lead to cell swelling.
   2. Alternatively, if the cell swells, transport mechanisms that extrude solutes out of the cell (eg, **K^+ or Cl^- channels or the K^+-Cl^2 cotransporter**) will be activated.
   3. Because of the transport mechanisms involved, cell volume regulation depends ultimately on the Na^+ and K^+ ionic gradients generated by the **Na^+/K^+ pump.**

J. **Regulation of cellular pH** at a constant level is critical for cell function.
   1. Changes in cellular pH can alter the conformation of proteins with ionizable groups (including a variety of enzymes and channels), thus affecting their function.
   2. Transport mechanisms that carry either H^+ or HCO_3^- (bicarbonate) are important for the maintenance of cellular pH. Transporters include the **Na^+-H^+ exchanger,** which alkalinizes the cytosol, and the **K^+-H^+ exchanger** in corneal epithelium, which acidifies the cytoplasm.

K. **Epithelia** are sheets of specialized cells that link the body to the external environment.
   1. Epithelia are **polarized** at the structural, biochemical, and functional levels. This means that one side of the epithelial sheet contains different electrochemical gradients across its **apical** and **basolateral membranes.**
   2. **Transepithelial transport** can be in the form of either secretion or absorption. Solutes can cross an epithelial cell layer by moving through the cells (**transcellular pathway**) or by moving between cells (**paracellular pathway**). Epithelia are classified as tight or leaky based on the permeability of their (paracellular) tight junctions to ions.
   3. To understand how **absorption** through an epithelial cell layer occurs, consider the example of a NaCl-absorbing epithelium in the small intestine.
      a. The primary Na^+ entry pathway is on the apical side and varies with the tissue. It can be either a **Na^+ channel** or a transporter such as the Na^+-H^+ exchanger or Na^+-coupled cotransporters (eg, Na-glucose, Na–amino acid). Na^+ channels on the apical membrane are members of the **amiloride-sensitive Na^+-channel** family.
      b. **Na^+ efflux** across the basolateral membrane is performed by the Na^+/K^+ pump. Therefore, Na^+ enters at the apical side and is secreted at the basolateral side, resulting in net transport of Na^+ across the epithelium.
      c. **Cl^- follows Na^+** movement across the epithelium through either the transcellular or the paracellular pathway, depending on the tissue.

(1) The **transcellular pathway** refers to ion **movement through the cell layer,** whereas the **paracellular pathway** refers to **ion movement between cells.**

(2) The driving force for Cl^- movement through the paracellular pathway is the electrical potential generated by the net movement of Na^+ (positive on the basolateral side).

(3) Alternatively, if Cl^- crosses the epithelium through the transcellular pathway, it usually enters at the apical side through transporters (eg, **Cl^--HCO_3^{2} exchanger, Na^+-K^+-$2Cl^-$ cotransporter**) and leaves the cell at the basolateral side through Cl^- channels or the K^+-Cl^- cotransporter.

d. **The activity of the Na^+/K^+-ATPase** on the basolateral side will result in the transport of K^+ ions inside the cell. Therefore, to maintain steady-state ion concentration in the cytosol, the cell must have a mechanism to recycle the pumped K^+. This mechanism involves a variety of K^+ channels located on the basolateral membrane.

4. **Secretion** is conceptually more difficult than absorption, but the same principles discussed for absorption apply.
   a. The Na^+/K^+-ATPase on the basolateral membrane pumps Na^+ out and K^+ into the cell. K^+ is recycled back into the extracellular fluid through the action of K^+ channels on the basolateral membrane.
   b. The Na^+ gradient generated by the Na^+/K^+-ATPase is used to drive the Na^+-K^+-$2Cl^-$ (or K^+-Cl^-) cotransporter on the basolateral membrane, resulting in the net transport of Cl^- into the cell.
   c. The increased Cl^- concentration inside the cell causes Cl^- secretion through Cl^- channels on the apical membrane, resulting in net Cl^- transport across the epithelial cell layer.
   d. The combined secretion of Cl^- into the lumen (apical side) and efflux of K^+ through K^+ channels on the basolateral membrane results in a transepithelial potential that is more negative on the luminal side. This negative potential drives the movement of Na^+ through the paracellular pathway toward the lumen.
5. Thus, epithelial cells can absorb or secrete solutes by inserting specific channels or transporters at either the apical or basolateral membrane.

L. **Intracellular calcium regulation** plays a physiologically important signaling and regulator role in various cellular processes. Cells have developed elaborate mechanisms to control Ca^{2+} levels and signals.

1. **Ca^{2+} signaling** in the cytoplasm occurs through a rise in Ca^{2+} levels, which activate Ca^{2+}-binding proteins that transduce the Ca^{2+} signal into a cellular response. Therefore, **maintenance of low cytoplasmic Ca^{2+} levels is required** for Ca^{2+} signaling.
2. A 20,000-fold concentration gradient exists for Ca^{2+} across the plasma membrane. Furthermore, cells also contain intracellular Ca^{2+} stores that are sequestered in the ER, which contains high levels of Ca^{2+}. Ca^{2+} signaling occurs through a rise in cytoplasmic Ca^{2+} levels due to either Ca^{2+} release from the ER or Ca^{2+} influx from the extracellular space.
3. Cells maintain low cytoplasmic Ca^{2+} levels by extruding Ca^{2+} out of the cell using the plasma membrane Ca^{2+}-ATPase and the Na^+-Ca^{2+} exchanger, or by sequestering Ca^{2+} into the ER using the ER Ca^{2+}-ATPase.
4. **Cells increase** their cytoplasmic **Ca^{2+} levels in response to primary signals** such as hormones and growth factors.

a. Once the primary signal is received, Ca^{2+} channels on the ER membrane or in the cytosol open, releasing Ca^{2+} into the cytoplasm and transducing the primary signal into a cellular response.
b. Channels on the ER membrane that mediate Ca^{2+} release include the inositol 1,4,5-triphosphate (**IP_3**) receptor and the **ryanodine receptor.**
c. Ca^{2+} influx from the extracellular space is mediated by different channel classes, including ligand-gated channels (such as the **AchR**) and **voltage-gated channels** (such as the Ca^{2+} channels in cardiac muscle).

DISEASES ASSOCIATED WITH CALCIUM REGULATORY DEFECTS

- ***Malignant hyperthermia*** *is a genetic disorder where affected individuals react abnormally to volatile anesthetics, particularly* ***halothane,*** *and muscle relaxants such as* ***carbachol.***
  –Malignant hyperthermia is ***due to mutations in the ryanodine receptor*** *leading to an overactive receptor. The mutated ryanodine receptor is especially sensitive to the aforementioned anesthetics, resulting in increased Ca^{2+} release and hyperthermia as well as sustained muscle contraction (rigidity).*
  –Left untreated, respiratory and lactic acidosis, and extensive necrosis of muscle cells follow, leading to hyperkalemia, cardiac arrhythmias, and often-lethal ventricular fibrillation.
  –High Ca^{2+} levels will also lead to the continuous activation of the ER Ca^{2+}-ATPase and ATP hydrolysis, resulting in increased heat production and hyperthermia.
  *–****Vigorous exercise*** *could also lead to abnormal muscle contraction in individuals with malignant hyperthermia.*
  –Therapy involves ***treatment with dantrolene,*** *which inhibits the ryanodine receptor and the uncontrolled muscle contraction.*
- ***Brody disease*** *is an autosomal recessive mutation in the ER Ca^{2+}-ATPase, which leads to* ***exercise-induced impairment of skeletal muscle relaxation.***
- ***Darier disease*** *is a skin disorder* ***due to mutations in the ER Ca^{2+}-ATPase,*** *leading to disruption of the cytoskeleton of skin cells and loss of adhesion between these cells.*
- ***X-linked congenital stationary night blindness*** *is a recessive disease of the human retina* ***due to mutations in a voltage-gated Ca^{2+} channel,*** *leading to defects in glutamate release and neurotransmission, which impairs the function of rod and cone cells in the retina.*
- ***Lambert-Eaton myasthenic syndrome (LEMS)*** *is an autoimmune disease characterized by an* ***increased number of LEMS antibodies*** *against presynaptic Ca^{2+} channels, leading to defective neurotransmission and weakness of limb muscles. Repeated stimulation of affected muscles leads to increased action potentials and muscle strength.*

III. Cell Signaling

A. Types of Cell Signaling

1. **Autocrine signaling** involves a secreted **substance acting on** the same **cell that produced it. Examples include amino acids, steroids, and polypeptides.**
2. **Paracrine signaling** involves a **substance diffusing** from the signaling cell that produced it **to nearby target cells** to elicit a response. For example, the gastrointestinal regulatory peptide somatostatin is produced by D cells in the stomach and diffuses to gastric acid cells to decrease secretion.
3. **Endocrine signaling** involves a **substance** secreted by endocrine cells that is **transported** in the blood **to distant target cells** to elicit a response. For example, adrenocorticotropic hormone, which is released from the anterior pituitary into the blood, stimulates the release of cortisol from the adrenal cortex.
4. Neurocrine signaling involves the release of neurotransmitters at synaptic junctions from nerve cells that act on postsynaptic cells.

B. Cell Signaling Events

1. A signaling cell produces a signaling molecule termed a **ligand** or **primary messenger,** which binds a receptor associated with a target cell.
2. Ligand binding results in **conformational change and activation of the receptor.**
3. The activated receptor elicits a response in the target cell, either directly or indirectly through the production of a secondary signal termed a **second messenger.**
   a. Target cell responses include alterations in cellular metabolism and alterations in gene transcription.
   b. Second messenger **examples include cAMP, DAG** (diacylglycerol), and **IP_3.**
   c. Hormone binding to a **G-protein** results in activation of **phospholipase C,** which catalyzes **phosphatidylinositol 4,5-diphosphate** to form IP_3 and DAG.

C. Types of Receptor Classes

1. **Intracellular receptors** located in the cytoplasm or nucleus of the target cell are bound by **lipophilic ligands,** which diffuse through the membrane of the target cell.
   a. Ligand binding alters the receptor's conformation, exposing the receptor's **DNA-binding domain.**
   b. Receptors bind specific gene promoter elements and activate transcription of specific genes that results in the synthesis of specific proteins.
   c. An example is an **estrogen receptor** in uterine smooth muscle cells.
2. There are four types of **cell surface receptors** (Figure 1–2):
   a. **Nicotinic cholinergic receptors** are linked to **ligand-gated ion channels** that are selectively permeable to specific anions or cations (eg, nicotinic AchRs on muscle cells).
   b. **Catalytic receptors** are transmembrane proteins that have intrinsic enzymatic (eg, **serine or tyrosine kinase**) activity.
   c. Other receptors are **linked to proteins with enzymatic activity.**
      (1) These receptors do not have catalytic activity themselves.
      (2) An example is cytokine receptor signaling through cytoplasmic tyrosine kinase (eg, the **JAK/TYK-STAT system**).
   d. **G-protein-linked receptors** have an extracellular ligand-binding domain and an intracellular domain that binds G-proteins (Figure 1–3).
      (1) After ligand binding, the receptors interact with G-proteins.
      (2) G-proteins are heterodimeric, **consisting of α, β, and γ subunits** that dissociate.
      (3) G-proteins (α-subunits) bound to GTP interact with and activate specific membrane-bound enzymes, resulting in the production of second messengers that elicit responses in target cells.
      (4) An example is an **adenylate cyclase system.**

CELL SIGNALING ERROR–INDUCED DISEASE

- ***Cholera***
  –***Cholera toxin*** *alters G-protein so that* ***guanosine triphosphatase*** *(GTPase) is unable to hydrolyze GTP, resulting in prolonged stimulation of adenylyl cyclase and increased production of cAMP.*
  –*Elevated cAMP in intestinal epithelial cells results in massive gut secretion of water and electrolytes, resulting in severe diarrhea and dehydration.*

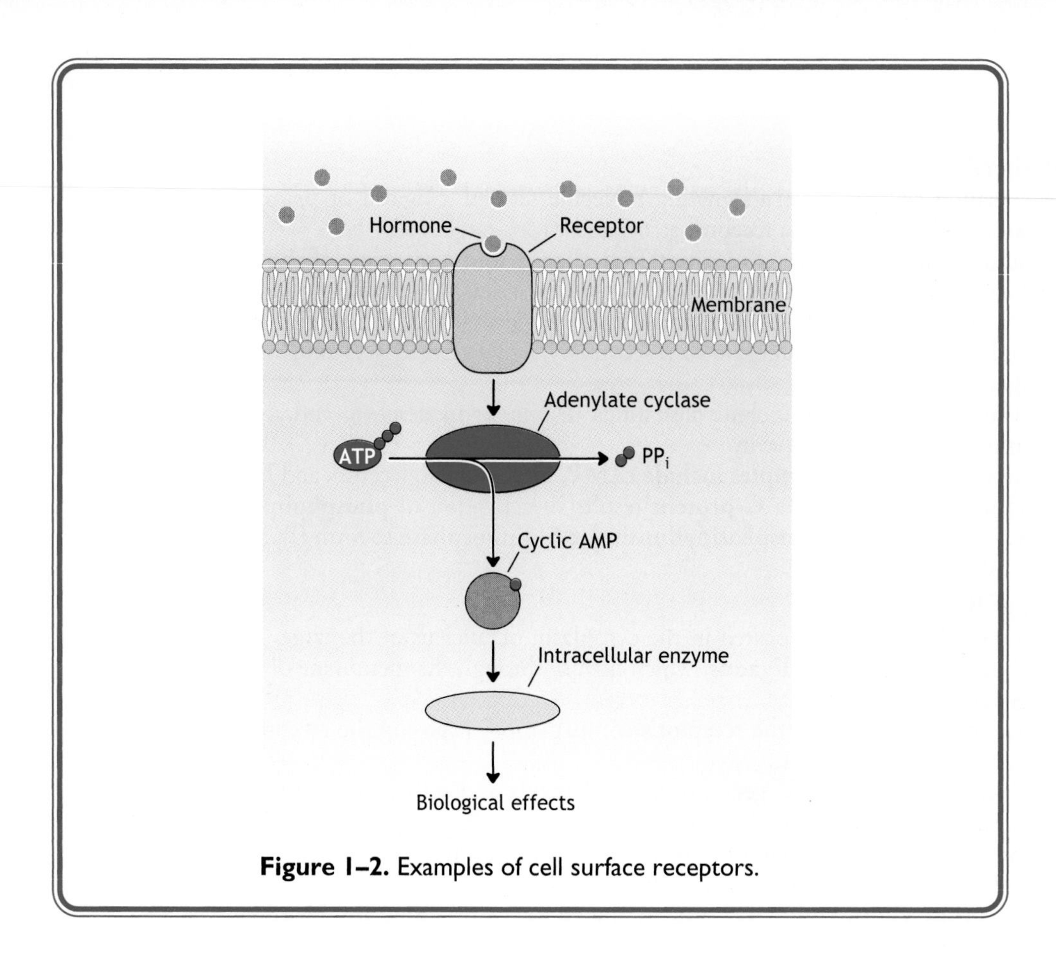

Figure 1–2. Examples of cell surface receptors.

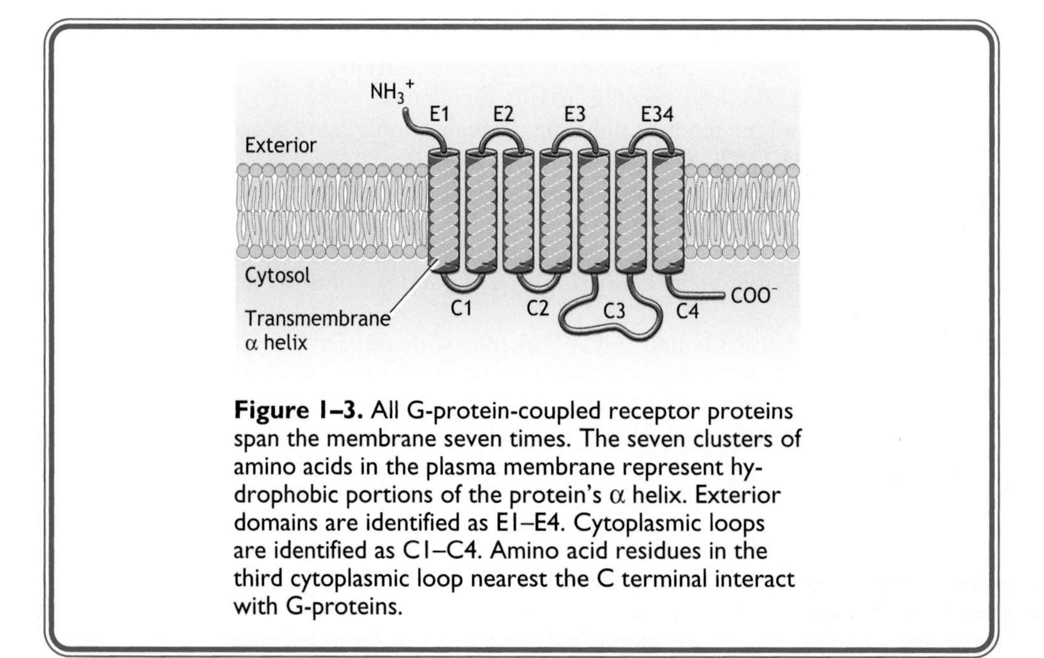

Figure 1–3. All G-protein-coupled receptor proteins span the membrane seven times. The seven clusters of amino acids in the plasma membrane represent hydrophobic portions of the protein's α helix. Exterior domains are identified as E1–E4. Cytoplasmic loops are identified as C1–C4. Amino acid residues in the third cytoplasmic loop nearest the C terminal interact with G-proteins.

- ***Pseudohypoparathyroidism***
  –Pseudohypoparathyroidism results from a defective G-protein and causes decreased cAMP levels.
  –Patients exhibit symptoms of hypoparathyroidism with normal or slightly elevated parathyroid hormone levels.
- ***Pertussis** (**Whooping Cough**)*
  –Pertussis toxin blocks the activity of G1, allowing adenylate cyclase to stay active and increase cAMP.

IV. Membrane Potential

A. The **membrane potential** is the difference in electrical potential (voltage) between the inside and outside membrane surfaces under resting conditions.

B. Cells have an excess of negative charges at the inside surface of the cell membrane and exhibit a negative membrane potential at rest.

1. Because the K^+ concentration inside the cell is higher than the outside concentration, K^+ moves out of the cell, leaving excess negative charges on the inside of the cell membrane.
2. The Na^+/K^+ pump acts as a second factor to generate negative charges on the inner membrane surface by pumping three Na^+ out and only two K^+ in.
3. The K^+ efflux is primarily responsible for the resting membrane potential.

C. The **equilibrium potential** is the membrane potential that exists if the cell membrane becomes selectively permeable for an ion, causing the distribution of the ion across the membrane to be at equilibrium.

1. The **Nernst equation** describes the relationship between the concentration gradient of an ion and its equilibrium potential. Thus, the magnitude of the equilibrium potential can be calculated by the Nernst equation:

$$E = \frac{RT}{FZ} \, In \, \frac{Co}{Ci}$$

where
E = equilibrium potential (volts)
R = the gas constant
T = the absolute temperature
F = Faraday's constant (2.3×10^4 cal/V/mol)
Z = the valence of the ion (+1 for Na^+, +2 for Ca^{2+})
In = logarithm to the base 10
Co = the outside concentration of the positively charged ion
Ci = the inside concentration of the positively charged ion

2. In spinal nerve cells the resting membrane potential is –70 mV, which is near the K^+ equilibrium potential of –90mV. Therefore, nerve cell membranes are selectively permeable to K^+.
3. The Nernst equation predicts that the equilibrium potential for K^+ will be negative because K_0 is less than K_i. It also predicts that the equilibrium potential for Na^+ will be positive because Na_0 is greater than Na_i.
4. Because the membrane is most permeable to K^+ and Cl^-, the actual membrane potential of most cells is around –70 mV.

D. **Resting membrane potential** is the potential difference across the cell membrane in millivolts (mV).

1. The resting membrane potential is **established by different permeabilities or conductances** of permeable ions.
   a. For example, the resting membrane potential of nerve cells is more permeable to K^+ than to Na^+.
   b. Changes in ion conductance alter currents, which change the membrane potential.
   c. **Hyperpolarization** is an increase in membrane potential in which the inside of the cell becomes more negative.
   d. **Depolarization** is a decrease in membrane potential in which the inside of the cell becomes more positive.
2. An **action potential** is a rapid, large decrease in membrane potential (ie, depolarization) (Figure 1–4).
   a. Action potentials usually occur because of increases in the conductance of Na^+, Ca^{2+}, and K^+ ions.
   b. The **threshold** is the membrane potential that induces an increase in Na^+ conductance to produce an action potential.
   c. **Depolarization** produces an opening of the Na^+ channel through fast opening of the activation gates and slow closing of the inactivation gates.
   d. **Closure of the inactivation gates** results in closure of the Na^+ channels and decreased Na^+ conductance.
   e. Slow opening of the K^+ channels increases K^+ conductance higher than Na^+ conductance, resulting in repolarization of the membrane potential.
   f. Thus, **repolarization** is the return of the membrane potential to its original value due to an outward K^+ movement.
3. The **refractory period** is the period during which the cell is resistant to a second action potential.
4. During the **relative refractory period** only some of the inactivated Na^+ channels are reset and K^+ channels are still open. Thus, another action potential can be elicited if the stimulus is large enough.

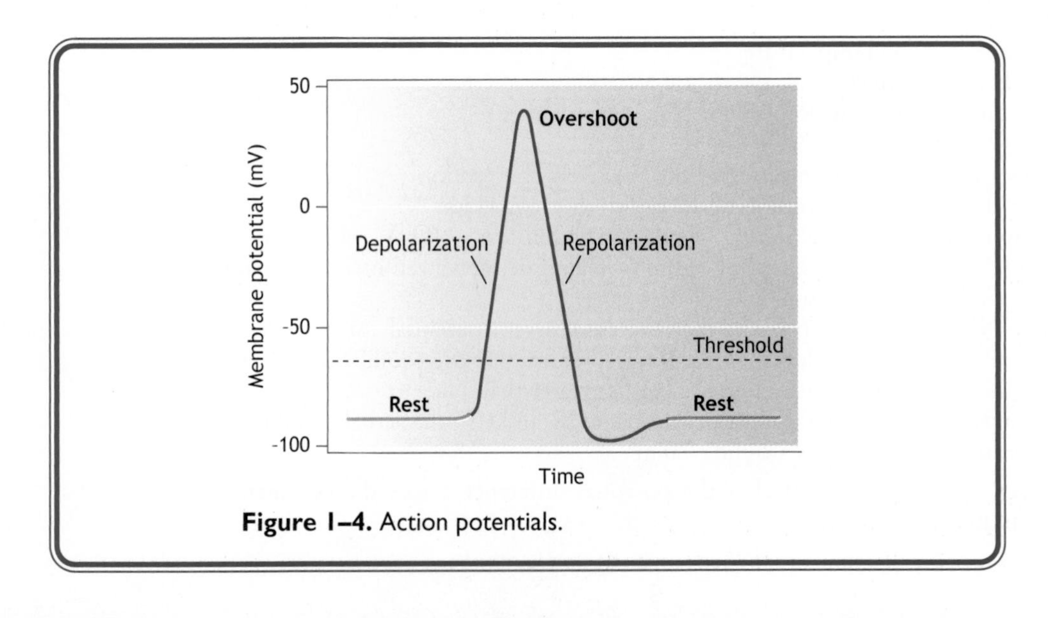

Figure 1–4. Action potentials.

5. Propagation of the action potential requires a system that regenerates the action potential along the axon.
   a. **Conduction velocity** is **increased by** increased fiber size and **myelination** and is dependent on the magnitude of the depolarizing current.
   b. Myelinated nerves exhibit saltatory conduction in which the action potential skips from node to node where the voltage-gated Na^+ channels congregate.
6. **Depolarization block** occurs when a depolarization stimulus occurs slowly so that Na^+ channels may inactivate before enough Na^+ channel openings occur. Thus, even though the membrane potential exceeds the threshold, no action potential is produced.
7. **Organophosphate poisoning** occurs by depolarization block of neuromuscular junctions, thereby inhibiting acetylcholine esterase (AchE) from breaking apart acetylcholine molecules.

V. Structure of Skeletal Muscle

A. **Skeletal muscle** is organized into progressively smaller anatomical units and contracts in response to neuromuscular synaptic transmission.

B. Muscle fibers are surrounded by a plasma membrane more commonly called the **sarcolemma.**

C. Muscle fibers are composed of a bundle of fibrous structures called **myofibrils,** and each myofibril is a linear arrangement of repeating structures called **sarcomeres** that consist of smaller filaments called **myofilaments.**

D. Sarcomeres are the fundamental contractile unit of skeletal muscle and are characterized by their highly ordered appearance under a polarizing-light microscope (Figure 1–5).
1. **Thick filaments** in the A band are composed primarily of the protein **myosin.**
   a. Each myosin molecule is composed of six monomers: two protein strands intertwined in a helical arrangement (termed **heavy chains**) and four smaller, globular proteins (termed **myosin light chains**). There are two essential light chains and two myosin regulatory light chains.
   b. Each heavy chain is associated with a **globular head.** The two globular heads of myosin heavy chains can hydrolyze ATP to ADP and inorganic phosphate and also have the intrinsic ability to interact with actin.
   c. The **rod-like region** (or tail) stabilizes the protein and tends to self-aggregate spontaneously, thereby forming the thick filament.
   d. Treatment with the proteolytic enzyme trypsin splits myosin into two components, **heavy meromyosin** and **light meromyosin.** Another proteolytic enzyme, **papain,** cleaves heavy meromyosin into a globular protein, $\mathbf{S_1}$, and a rod-like protein, $\mathbf{S_2}$.
   e. The sites sensitive to proteolytic digestion are regions that allow flexing of the molecule, also called **hinge regions.**
2. **Thin filaments** are composed of three primary proteins: **actin, tropomyosin,** and **troponin.**
   a. Actin can exist in two states: globular **G-actin** and filamentous **F-actin.**
   b. G-actin polymerizes to form F-actin.
   c. Each G-actin monomer contains binding sites for myosin, tropomyosin, and troponin I.
   d. The basic structure of the thin filament consists of two strands of intertwined F-actin in a double helical arrangement.

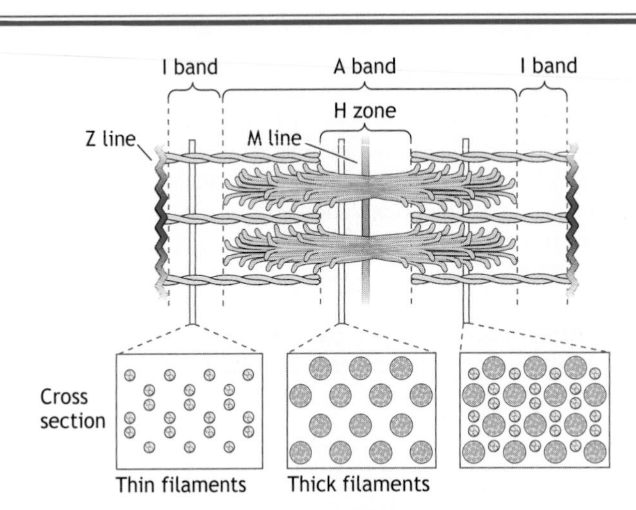

Figure 1–5. Sarcomere structure. The A bands contain the thick filaments. The I bands contain the thin filaments, which are attached to and extend from the Z line. The Z line maintains the regular spacing of the thin filaments within the sarcomere. The space between terminations of thin filaments is called the H zone, and the denser area within the H zone is termed the M line.

e. Tropomyosin is an elongated protein that lies within the two grooves formed by the double stranded F-actin (Figure 1–6).

f. Each thin filament contains 40–60 tropomyosin molecules.

g. Troponin is a complex of three separate proteins:

(1) **Troponin T** binds the other two troponin subunits to tropomyosin.

(2) **Troponin C** binds Ca^{2+}, the crucial regulatory step in muscle contraction, and is closely related to another Ca^{2+} binding protein.

(3) Each troponin C molecule in skeletal muscle has two high-affinity Ca^{2+} binding sites that participate in binding of troponin C to the thin filament.

(4) **Troponin I** is responsible for the inhibitory conformation of the tropomyosin-troponin complex observed in the absence of Ca^{2+}. Troponin I binds to actin and prevents contraction.

3. **Tubules,** a tubular network, are located at the junctions of A bands and I bands and contain a protein called the **dihydropyridine receptor.**
4. The **sarcoplasmic reticulum** (SR) is the site of Ca^{2+} storage near the **transverse tubules** (T-tubules). It contains a Ca^{2+}-release channel known as the **ryanodine receptor.**

E. Several steps are involved in the **mechanics of muscle contraction:**

1. Action potentials in muscle cell membrane cause depolarization of the T-tubules, which opens Ca^{2+}-release channels in the SR and increases intracellular Ca^{2+}.
2. Ca^{2+} removes the troponin-tropomyosin inhibitory influence so that the active sites on each G-actin monomer are uncovered.

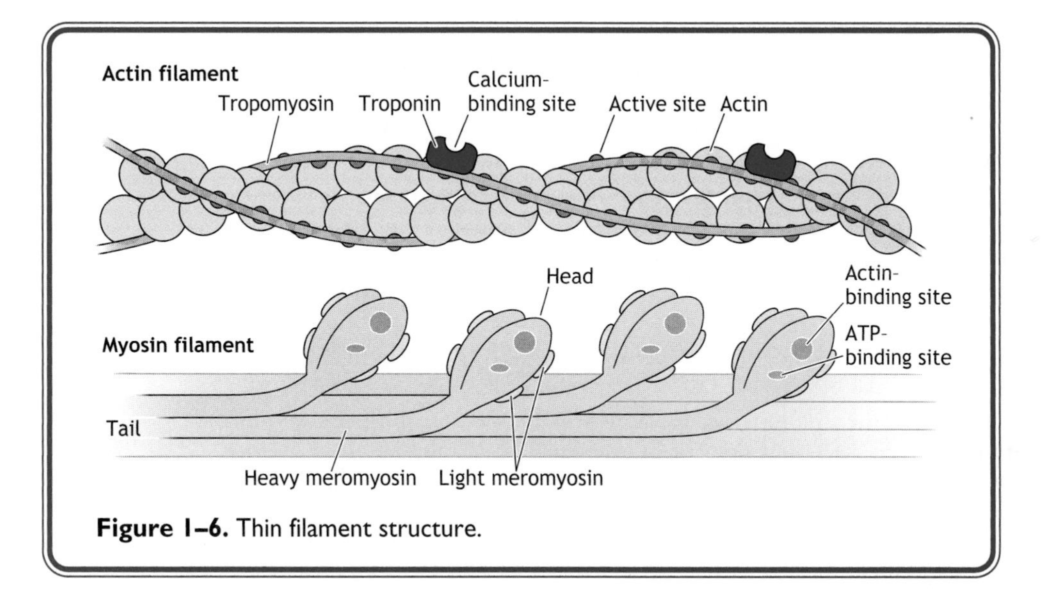

Figure 1–6. Thin filament structure.

3. The myosin globular heads that protrude from the thick filament bind with G-actin active sites, thus forming **crossbridges.**
4. **Intramolecular forces** (stored energy) within the myosin molecules allow myosin to flex in the so-called **hinge regions.** These areas are the two proteolytic enzyme–sensitive regions in the myosin molecule. The action of flexing of the myosin molecule causes the globular heads (still attached to actin) to tilt toward the center of the sarcomere. This movement, called the **power stroke,** creates tension that results from shortening of individual sarcomeres.
5. Immediately after the tilt, the crossbridge is broken and the globular heads snap back to the upright position.
6. At this point, a new crossbridge can be formed if ATP and Ca^{2+} are available in the vicinity of thick and thin filaments. In the absence of Ca^{2+}, crossbridge formation is not possible.
7. Relaxation occurs when Ca^{2+} uptake into the SR lowers intracellular Ca^{2+}.

F. **Cardiac Muscle**

The regulatory mechanism within cardiac muscle is similar to that of skeletal muscle except that troponin C from cardiac muscle has just one active low-affinity Ca^{2+} binding site.

G. The **biochemical events** that occur **during a skeletal muscle contraction cycle** involve an active complex and the rigor complex.

1. Myosin with ATP bound to it (myosin-ATP complex) has a low affinity for the G-actin active sites. When Ca^{2+} binds to troponin and tropomyosin, tropomyosin rotates out of the way so that the active sites on G-actin are uncovered. Myosin-ATP is simultaneously hydrolyzed to myosin-ADP, which has a high affinity for the G-actin active sites. Consequently, an **active complex, or crossbridge, is formed** between actin and myosin-ADP.
2. ADP is released from myosin, and the globular heads tilt toward the center of the sarcomere, producing tension. At this stage, the **rigor complex is formed** between actin and myosin.

3. ATP then binds to myosin, and the myosin-ATP complex breaks the crossbridge and the globular heads snap back to the upright position.
4. The cycle is ready to start again in the presence of Ca^{2+}.
5. Because ATP stores are small, the cell must regenerate the ATP needed for muscle contraction.

G. Skeletal muscle enters a state of prolonged stiffness termed **rigor mortis** at death.
1. Rigor mortis occurs because, with death, muscle cells are no longer able to synthesize ATP.
2. In the absence of ATP, the crossbridges between myosin and actin are unable to dissociate.
3. After 15–25 hours, proteolytic enzymes released from lysosomes begin to break down actin and myosin.

H. **Muscle length influences tension development by determining the amount of overlap between actin and myosin filaments.**
1. In an **isometric contraction,** muscle length is constant during the development of force. Thus, stimulation causes an increase in tension but no shortening. An example would be an individual pushing against an immovable object such as the wall of a house.
2. In an **isotonic contraction,** the muscle shortens while exerting a constant force or load. An example would be an individual lifting a glass of water to his or her mouth.
3. The tension that a stimulated muscle develops when it contracts isometrically is **total tension.**
4. The tension measured before muscle contraction is known as **passive tension.** The difference between the two values is the tension produced by the contractile process, the **active tension** (Figure 1–7).
5. If the muscle is stimulated to contract at any fixed length, an **active tension** develops because of crossbridge cycling. **The amount of active tension** developed is proportional to the number of crossbridges formed.
6. Tension is reduced when the sarcomere is shortened to a point where thin filaments overlap and prevent one another from forming crossbridges with myosin.
7. Thus, **isometric tension** produced **depends on the degree of overlap of the thick and thin filaments,** which dictates the number of crossbridges that can be formed.

I. The **force-velocity relationship** refers to the relationship between the load (or weight) placed on a muscle and the velocity at which that muscle contracts while lifting the load.
1. **Velocity** is the distance an object moves per unit time. A load can be thought of as a weight that the muscle is attempting to move via an isotonic contraction, for example, when a weightlifter tries to lift a series of progressively heavier weights.
2. A muscle can contract most rapidly with no load. At higher loads, however, the velocity of shortening is lower because more crossbridges are simultaneously active.
3. When the weight equals the maximum amount of force that the muscle can generate, the velocity becomes zero. In this case the contraction becomes isometric (eg, the muscle contracts but does not shorten).

J. The functional unit of a muscle is a group of muscle cells innervated by a single motor neuron and is called a **motor unit.**

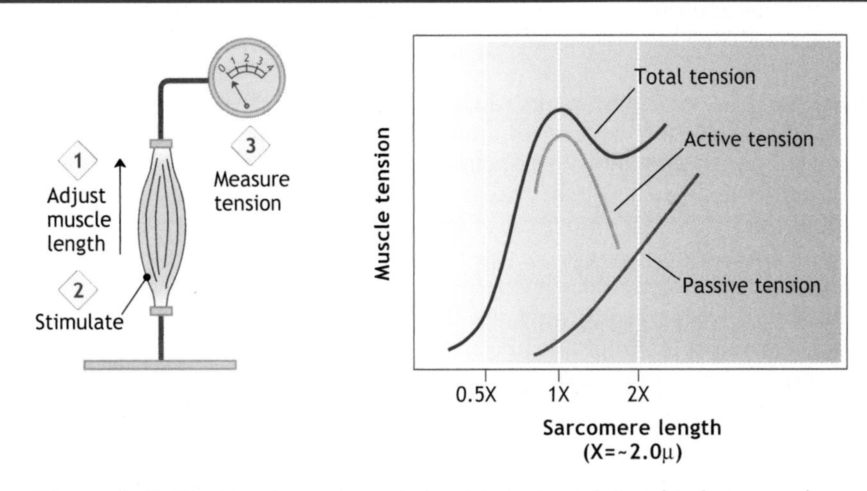

Figure 1–7. The length-tension relationship is the relationship between the length of the muscle and the amount of active or passive tension on the muscle. Active tension refers to the tension generated by the contractile forces when the muscle is stimulated, whereas passive tension refers to the elastic force acting on the muscle when the muscle is stretched. Total tension on the muscle is the sum of the active and passive tensions.

1. A motor unit consists of one **motor neuron,** its axon, and all the muscle cells innervated by that motor neuron. In adults, each muscle fiber is innervated by a **single motor axon.**
2. In general, **motor units in small muscles** that react to stimulation rapidly and subserve functions that require fine control have a low number of muscle fibers. An example is **laryngeal muscle,** in which a motor unit has approximately 2–3 muscle fibers per motor neuron.
3. **Motor units in large muscles** that subserve functions not requiring fine motor control tend to have a larger number of muscle fibers. An example is the **gastrocnemius,** in which a motor unit contains approximately 500 muscle fibers per motor neuron.
4. Because all the muscle cells in a motor unit contract together, the fundamental unit of contraction of a whole muscle is the contraction produced by a motor unit.
5. Increased tension development in skeletal muscle is attained by
   a. **Wave summation** (eg, increasing stimulus frequency of a single motor neuron).
   b. Summation, or **recruitment,** of motor units. Besides increasing tension development, recruitment allows a movement to be continuous and smooth because different motor units fire **asynchronously;** that is, while one motor unit is contracting, another might be at rest.

K. A contraction can be a single, brief contraction or a maintained contraction due to continuous excitation of muscle fibers.
   1. A single contractile event (eg, **twitch**) is initiated by a single action potential from a motor neuron reaching the neuromuscular junction.

2. If multiple stimuli are applied before the muscle fibers in the motor unit have relaxed, the force developed may be increased by summing multiple twitches in time.

a. This **summation** of contractions occurs when stimulation frequencies reach about 10 per second. As the frequency of stimulation is increased, the developed force continues to sum until a maximum developed force is reached.

b. At this point, the individual twitches occur so closely together that they fuse to produce a single smooth curve called **tetanus** (Figure 1–8).

c. Tetanus occurs in **skeletal muscle** because the **refractory period** (ie, the time during which the tissue does not respond to a second stimulus) **is short relative to the contraction time** and the force may be increased by summing the contractions of multiple fibers.

d. In cardiac muscle, **increase in the entry of Ca2+ enhances the contractile force.**

VI. Neuromuscular and Synaptic Transmission

A. The activity of various **skeletal muscle groups** is **controlled by the central nervous system** through innervation of individual muscle fibers.

B. Each motor nerve sends processes to each muscle fiber in the motor unit.

C. Where a motor nerve comes in contact with the surface of a muscle fiber, a highly organized and specialized structure is formed known as a **neuromuscular junction,** or **motor endplate** (Figure 1–9).

D. The invagination of the **muscle fiber sarcolemma** forms the **synaptic trough.**

E. The space between the axon terminal and invaginated sarcolemma is called the **synaptic cleft.**

F. Schwann cells are usually seen in the vicinity of the motor endplate and may isolate the synaptic cleft from extracellular space.

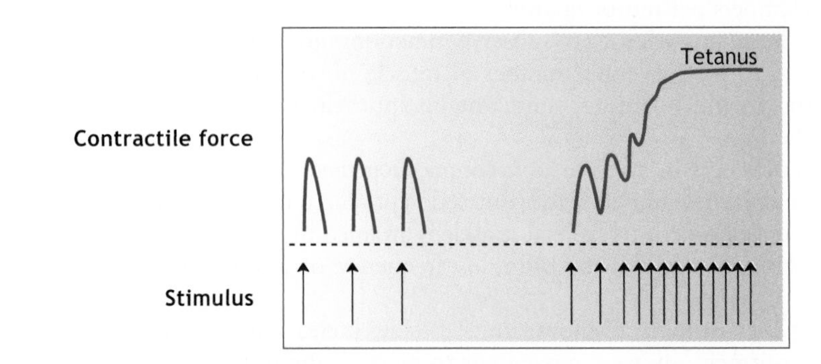

Figure 1–8. Recordings of contractile force during twitch contractions (*left*) and tetanic contraction (*right*) of skeletal muscle. A twitch contraction is a single brief muscle contraction that occurs in response to a single threshold stimulus. Tetanic contraction, or tetanus, is a constant contraction of skeletal muscle due to continuous excitation of muscle fibers.

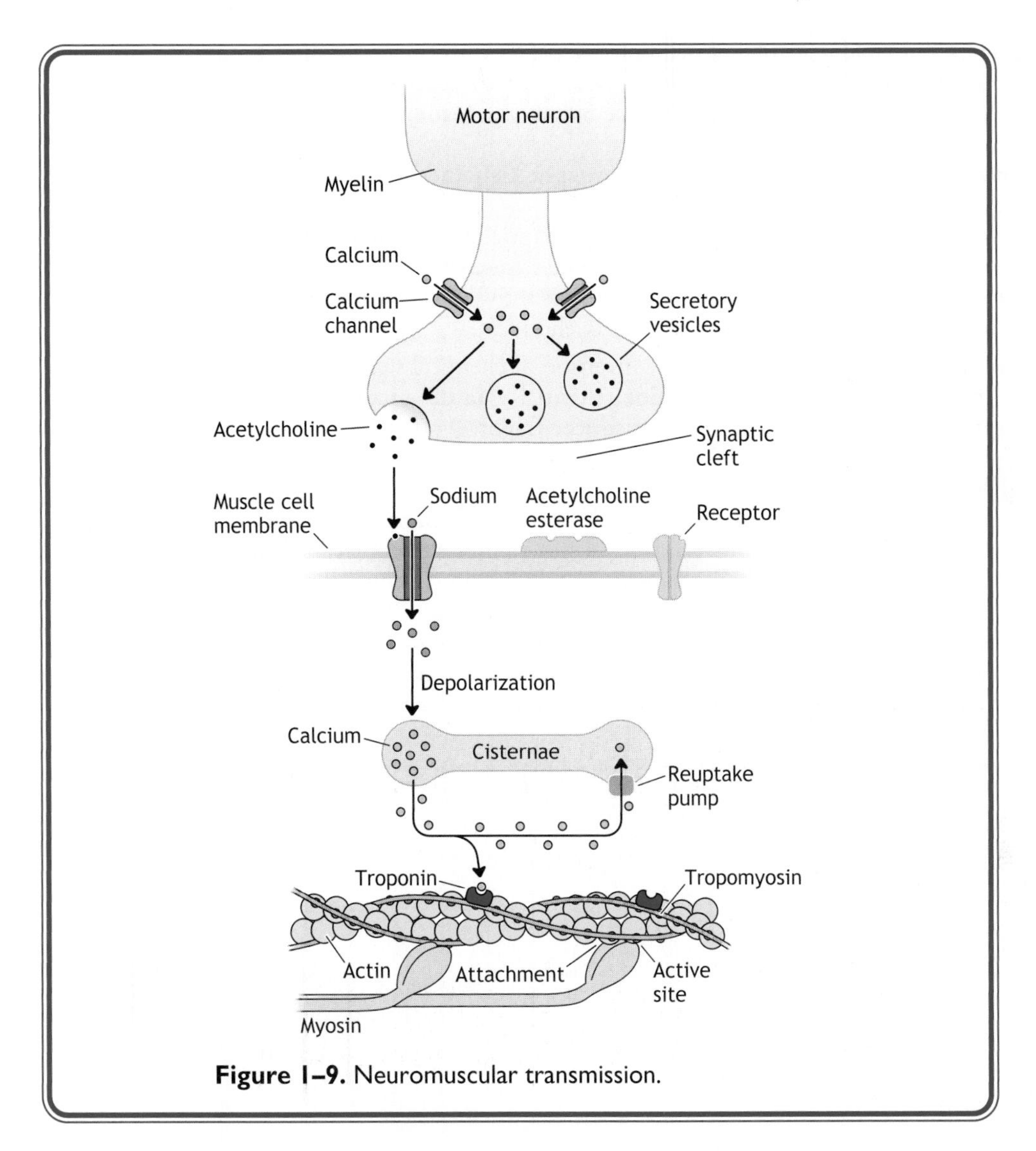

Figure 1–9. Neuromuscular transmission.

G. The neurotransmitter **acetylcholine is stored in synaptic vesicles** located in the axon terminal.

H. The **biosynthesis of acetylcholine** involves the reaction of choline with active acetate (acetyl-CoA).

1. The **key enzyme** in the biosynthesis of acetylcholine **is choline-*O*-acetyltransferase,** which is synthesized in the neuronal cell body and is transported to the axon terminal.
2. The **precursors** for the synthesis of acetylcholine **are pyruvate and choline.** Pyruvate is derived from the metabolism of glucose via glycolysis. Choline is actively taken up by the motor neuron.
3. Once synthesized, acetylcholine is **packaged into secretory vesicles** in the motor nerve terminal.

4. An action potential that reaches the motor nerve terminal increases the release of acetylcholine into the synaptic cleft. **Secretion of acetylcholine** involves fusion of the vesicles with the presynaptic membrane (**exocytosis**) and is triggered by a rise in Ca^{2+}.
5. Transmitter molecules diffuse across the synaptic cleft and bind to specific receptors on the postsynaptic cell.
6. Acetylcholine is rapidly **removed** from the synaptic cleft **via hydrolysis** into acetate and choline by the enzyme acetylcholinesterase (AchE).
7. Following hydrolysis of acetylcholine, choline is actively taken up by the nerve terminal and used for synthesis of new acetylcholine.

I. **Neuromuscular transmission** involves conversion of chemical signals (ie, acetylcholine) into electrical signals (ie, an action potential), **via the nicotinic AchR,** a ligand-gated ion channel that acts as a **transducer** (Figure 1–10).
1. The nicotinic AchR is an ionotropic receptor that is also an ion channel. Acetylcholine binding to the receptor opens the central core of the channel and increases the conductance of Na^+ and K^+ to move through the channel.
2. The entry of Na^+ causes **depolarization** of the membrane, which if of sufficient magnitude to reach **threshold,** produces an action potential that propagates over the entire surface of the muscle fiber (see Figure 1–10).
   a. **Crossbridge formation** between thick (myosin) and thin (actin) filaments depends on spreading of the action potential from the sarcolemma across the muscle fiber via the T-tubule system and subsequent release of Ca^{2+} from the SR.

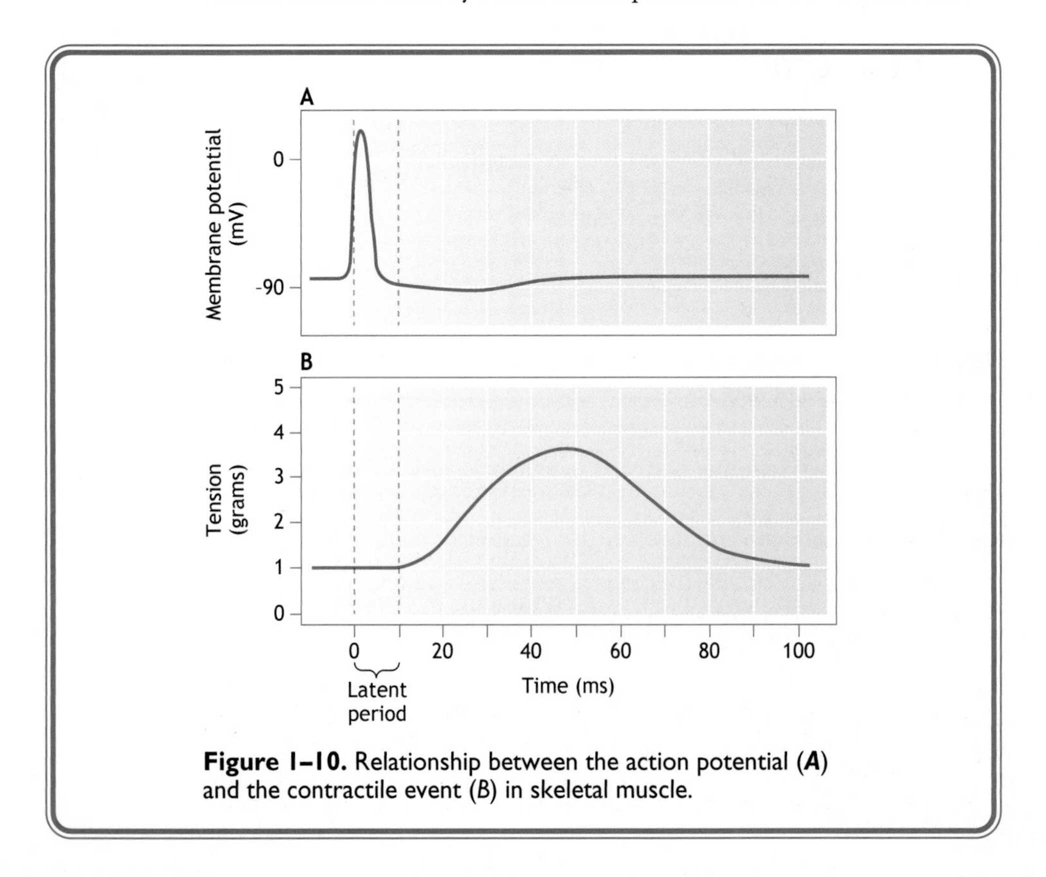

Figure 1–10. Relationship between the action potential (***A***) and the contractile event (*B*) in skeletal muscle.

b. If the initial depolarization at the motor endplate does not reach threshold, then excitation-contraction coupling and muscle contraction do not occur.

3. The **resting membrane potential,** or **endplate potential,** of skeletal muscle is approximately −70 mV (the interior of a muscle fiber is negative with respect to the exterior).

J. **Excitation-contraction coupling** refers to a series of events beginning with a muscle fiber action potential (the excitation phase of excitation-contraction coupling) and culminating with crossbridge formation and muscle fiber shortening (the contraction phase of excitation-contraction coupling).

1. A time lag, known as the **latent period,** occurs between the initiation of the muscle fiber action potential and the beginning of the actual contractile event.
2. Initiation of contraction starts with an action potential that begins at the **motor endplate** and travels along the sarcolemma of the muscle fiber.
3. The **T-tubules,** a continuation of the sarcolemma, carry the action potential to the core of the muscle fiber.
4. Portions of the T-tubules are close to the **terminal cisternae** of the SR, forming a structure called a **triad.**
5. A Ca^{2+}-ATPase, or **calcium pump,** actively pumps calcium from the cytoplasm into the interior of the SR.
6. An action potential reaching a triad serves as the stimulus for the SR to release calcium into the cytoplasm to allow crossbridge formation and muscle shortening.
7. Contraction ceases as the calcium is rapidly pumped back into the SR.

PHARMACOLOGIC AGENTS AND TOXINS AFFECTING THE NEUROMUSCULAR JUNCTION

- ***Curare:*** *This term refers to a group of substances originally used by Amazon Indians to kill animals. Curare-like compounds bind with high affinity to the AchR, block binding of acetylcholine, and thereby cause skeletal muscle paralysis. In modern medicine, muscle relaxation during abdominal surgery is the primary clinical use of curare or curare-like drugs.*
- ***α-Bungarotoxin:*** *This protein was isolated from* ***cobra snake venom.*** *It binds irreversibly to the AchR, blocks binding of acetylcholine, and like curare causes skeletal muscle paralysis. Victims of cobra bites usually die of suffocation.*
- ***Botulinum toxin:*** *The toxin produced by Clostridium botulinum inhibits release of acetylcholine from the nerve terminal. Death results from respiratory failure. Clinically, botulinum toxin is used to treat* ***focal dystonias,*** *which are neuromuscular disorders characterized by involuntary and repetitive skeletal muscle contractions. Examples of such disorders include hemifacial spasms and writer's cramp. Local treatment with botulinum toxin produces a chemical denervation.*
- ***Black widow spider toxin:*** *This toxin causes clumping of* ***acetylcholine-containing vesicles,*** *which results in excessive release of acetylcholine into the synaptic cleft.*
- ***Neostigmine and physostigmine:*** *These drugs are* ***anticholinesterase agents.*** *Their principal action is to inhibit AchE; the net effect is to increase the concentration of acetylcholine in the synaptic cleft. Clinically, physostigmine is used to treat* ***glaucoma*** *and* ***myasthenia gravis.***
- ***Organophosphates:*** *This broad group of agents includes insecticides and so-called nerve gases. Organophosphates are extremely toxic due to their essentially irreversible inactivation of AchE.*
- ***Benzodiazepines*** *(eg,* ***diazepam):*** *These agents are central nervous system depressants that do not act directly on the neuromuscular junction. Their muscle-relaxing effect is due to a depressant effect in the* ***reticular formation*** *of the brainstem.*
- ***Dantrolene:*** *This muscle relaxant acts by direct action on excitation-contraction coupling, inhibiting* Ca^{2+} *release by the SR.*

MYASTHENIA GRAVIS

- *Myasthenia gravis is a* ***neuromuscular disease*** *characterized by* ***weakness and marked fatigability of skeletal muscle.***
- *It is caused by an acquired autoimmune response in which antibodies are directed against nicotinic acetylcholine receptors (AchRs) and block binding of acetylcholine to the receptors.*
- ***Diagnosis*** *of myasthenia gravis is made* ***by the edrophonium test,*** *in which the patient is given edrophonium, an anticholinesterase; improvement in muscular strength suggests the disease.*
- ***Treatment*** *is aimed at reducing the strength of the immunologic attack or increasing cholinergic activity in the synapse and includes the following:*
  *–**AchE inhibitors** increase the concentration of acetylcholine in the synaptic cleft. Excessive treatment with AchE inhibitors can cause skeletal muscle weakness via desensitization of nicotinic AchR and can lead to a **cholinergic crisis.***
  *–**Corticosteroids** suppress the immune system and thereby reduce the concentration of circulating anti-AchR antibodies.*
  *–**Immunosuppressant drug therapy,** such as **azathioprine** or, less commonly, **cyclosporine,** is used in patients with severe disease that do not respond well to corticosteroids.*
  *–**Removal of the thymus gland** also suppresses the immune system because the thymus gland plays a role in maturation of T cells. One drawback is that sustained improvement may not begin until months or even years after the surgery.*
  *–**Plasmapheresis** involves removing plasma from the patient and replacing it with a plasma substitute. The overall effect of plasmapheresis is to reduce the concentration of circulating anti-AchR antibodies.*

VII. Smooth Muscle

A. Structure of Smooth Muscle

1. The **cytoplasm** of a smooth muscle cell **is homogeneous** (with no visible striations) when viewed by light microscopy.
2. **Specialized contacts** between individual smooth muscle cells **have two functions:** in communication and as mechanical linkages.
   a. **Gap junctions** (**nexus**) are areas of close opposition (~2 nm) between plasma membranes of separate cells. Gap junctions serve as a low-resistance electrical coupling structure.
   b. **Attachment plaques** are characterized by a 10- to 30-nm gap between plasma membranes of adjacent cells. These structures may serve as anchor points for thin filaments.
3. Smooth muscle cells contain SR but in less abundant quantities compared to skeletal muscle. Like skeletal muscle SR, the smooth muscle counterpart accumulates and releases Ca^{2+}.
4. Smooth muscle does not have a T-tubule system. However, **surface vesicles called caveolae** in individual cells are thought to have an analogous **role in transmission of action potentials.**
5. Smooth muscle contraction may also occur independently of increases in Ca^{2+} concentration.

B. Physiology of Smooth Muscle

1. Smooth muscle is typically subdivided into two classes: **unitary,** or **visceral, smooth muscle;** and **multiunit smooth muscle.**
2. Both classes of smooth muscle share the following characteristics:
   a. Smooth muscle may contract in response to either neuromuscular synaptic transmission or electrical coupling.

b. Smooth muscle is capable of contractions that are slow in onset but are sustained for long periods of time with relatively little energy input required.
c. The motor innervation of smooth muscle is exclusively autonomic, either parasympathetic or sympathetic.
d. All smooth muscle exhibits a certain degree of intrinsic tone, or basal resting tension; contractions are superimposed on this tone.
e. Some smooth muscles contract without action potentials.

3. **Visceral smooth muscle** performs important functions in the vascular system, the airways of the lung, the gastrointestinal tract, and the genitourinary tract. The following **general characteristics** enable visceral smooth muscle to carry out these functions:
   a. Spontaneous electrical **activity is initiated in pacemaker areas** and spreads throughout the entire muscle. Unlike pacemakers in cardiac muscle, smooth muscle pacemakers move around.
   b. **Tension develops** in response to stretch.
   c. Generally, **contractions** are initiated by circulating hormones and are not typically initiated by motor nerve impulses. However, contractile activity may be modified and regulated by motor nerve input.
   d. Visceral smooth muscle is **widely distributed** in a variety of tissues and organs. Examples include the gastrointestinal tract, uterus, and arterioles.
   e. **Spontaneous activity** in visceral smooth muscle **results from** at least two types of fluctuations in electrical activity:
      (1) **Slow waves** of depolarization are produced when the threshold is reached, as occurs in longitudinal muscles of the intestines.
      (2) **Spontaneous prepotentials,** or **spike potentials,** produce an asynchronous discharge resulting in irregular contractions such as occurs in the nonpregnant uterus.
   f. Unlike skeletal muscle, smooth muscle can contract or relax in response to either neuronal or humoral stimulation.
   g. Calcium is the signal for contraction in smooth muscle and both extracellular and intracellular Ca^{2+} activates contraction.
   h. Because smooth muscle does not contain troponin, Ca^{2+} binds to **calmodulin** and then the **Ca^{2+}-calmodulin complex** activates the enzyme **myosin light chain kinase (MLCK)**.
   i. Ca^{2+}-calmodulin-activated MLCK phosphorylates the heavy meromyosin component of myosin and thereby greatly increases ATPase activity. Phosphorylated myosin has a high affinity for actin, and crossbridges form between myosin and actin.
   j. Remember that smooth muscle can maintain high force at a low rate of ATP hydrolysis.
   k. **Relaxation of smooth muscle** can occur through the following mechanism:
      (1) **Stimulation of Ca^{2+}-pumping activity** of either the plasma membrane or the SR reduces the concentration of Ca^{2+} in the vicinity of the contractile elements.
      (2) The **activity of myosin light chain phosphatase** can be increased to dephosphorylate the MLC to relax the smooth muscle.
   l. **Multiunit smooth muscle** is more similar to skeletal muscle than to visceral smooth muscle but is much less abundant than visceral smooth muscle.

(1) Multiunit smooth muscle does not contract spontaneously.
(2) Multiunit smooth muscle is usually activated by motor nerve stimulation. Multiunit smooth muscle is only minimally responsive to circulating hormones.
(3) Multiunit smooth muscle does not respond to stretch by developing tension.
(4) **Examples** of multiunit smooth muscle **include ciliary muscle** (the muscle that focuses the eye), **pilomotors** (the muscles that cause hair erection), and **nictitating membranes** (in the eyes of cats).

CLINICAL PROBLEMS

A 27-year-old woman presents with muscle weakness, including eyelid ptosis, slurred speech, and difficulty swallowing. The history shows that she is being treated for a gram-negative infection with gentamicin. The following tests have been ordered: thyroid function studies, serum creatine kinase, an electromyogram, and a muscle biopsy.

The attending physician chides the resident on the case for not ordering an edrophonium test, which produces a dramatic improvement in the woman's muscle strength when administered intravenously. All of the other tests returned with normal values.

1. The resident's working diagnosis is
   A. Duchenne muscular dystrophy
   B. Monoadenylate deaminase deficiency
   C. Myasthenia gravis
   D. Hyperthyroidism
   E. Toxic drug myopathy

2. This patient's condition most likely results from
   A. Inadequate acetylcholinesterase in the synaptic cleft
   B. Production of defective acetylcholine receptors
   C. Impaired synthesis or storage of acetylcholine in presynaptic vesicles
   D. Impaired release of acetylcholine from presynaptic terminals
   E. Blockade and increased turnover of acetylcholine receptors

Cholera toxin can affect cells by blocking the guanosine triphosphatase (GTPase) activity of their G_S-proteins.

3. On a cellular level, which one of the following would be helpful in reducing the harmful effect of cholera toxin?
   A. Increasing the amount of intracellular cyclic adenosine monophosphate (cAMP)
   B. Inhibiting the activity of the adenylate cyclase in the cell
   C. Inhibiting the G_i-protein within the cell

D. Adding ligand for the G_S-protein-linked receptor

E. Increasing the amount of protein kinase A in the cell

A 45-year-old woman experiences blurred vision and difficulty swallowing after eating some home-canned vegetables. These symptoms are followed by respiratory distress and flaccid paralysis.

4. The symptoms of her illness are most associated with which of the following?

   A. Black widow spider toxin

   B. Botulinum toxin

   C. Organophosphate poisoning

   D. Benzodiazepine ingestion

   E. α-Bungarotoxin

5. This toxin exerts its action by

   A. Binding irreversibly to the acetylcholine receptor to cause paralysis

   B. Causing a clumping of acetylcholine-containing vesicles, resulting in excessive release of acetylcholine into the synaptic cleft

   C. Inhibiting the release of acetylcholine from the nerve terminal

   D. Inhibiting anticholinesterase to increase the concentration of acetylcholine in the synaptic cleft

   E. Inhibiting the release of calcium from the sarcoplasmic reticulum

A 5-year-old boy has a history of growth retardation; pulmonary infections; and bulky, oily, malodorous stools.

6. Which of the following test results would be expected in this patient?

   A. Abnormal sweat chloride test

   B. Low C_3 complement level

   C. Abnormal nitroblue tetrazolium (NBT) dye test

   D. Positive wheel and flare reaction with antigen scratch testing

   E. Sputum with gram-positive diplococci

7. This disease is due to

   A. A direct blockade of sodium channels in the plasma membrane

   B. A reduced number of chloride channels on the cell membrane

   C. The direct blockade of the potassium channel gated by a G-protein

   D. A net increase in ion flex through the calcium channel, stimulating neurotransmitter secretion

   E. A blockade of ligand-gated ion channels in neuronal cell membranes

ANSWERS

1. The answer is C. Edrophonium is an anticholinesterase agent that improves muscle strength in myasthenic patients by increasing the acetylcholine concentration in the synaptic cleft. The test is diagnostic for myasthenia gravis. Duchenne muscular dystrophy (choice A) is a defect in the gene encoding dystrophin, a cytoskeletal protein. Patients with this disorder experience progressive muscle weakness. Adenosine deaminase deficiency (choice B) causes severe combined immunodeficiency with impaired T cell and B cell function. Hyperthyroidism (choice D) is characterized by palpitations, sweating, heat intolerance, functional muscle tremor, and exophthalmos, not by the symptoms described in this case. The edrophonium test differentiates myasthenia gravis from toxic drug myopathy (choice E).

2. The answer is E. Myasthenia gravis is a neuromuscular disorder resulting in muscle weakness. It is caused by an autoimmune response to acetylcholine receptors, leading to increased turnover and a reduced number of these receptors.

3. The answer is B. Cholera toxin causes a functional derangement of sodium and water transport in the gut. The toxin binds to the GM_1-ganglioside receptors of the luminal membrane of enterocytes and activates epithelial adenylate cyclase. Thus, inhibiting adenylate cyclase activity would reduce the harmful effects of cholera toxin.

4. The answer is B. Botulinum toxin inhibits the release of acetylcholine from the nerve terminal, resulting in blurred vision, ptosis, unreactive pupils, paralysis, and respiratory failure. Black widow spider toxin (choice A) causes clumping of acetylcholine-containing vesicles, resulting in excessive acetylcholine release into the synaptic cleft. Organophosphate poisoning (choice C) blocks acetylcholinesterase action, resulting in a massive cholinergic response. Benzodiazepines (choice D) induce muscle relaxation through the depression of the reticular formation in the central nervous system. α-Bungarotoxin (choice E) blocks the binding of acetylcholine to its receptor by irreversibly binding to the acetylcholine receptor.

5. The answer is C. Toxins produced by *Clostridium botulinum* cleave specific presynaptic proteins, preventing neurotransmitter release at both neuromuscular and parasympathetic cholinergic synapses.

6. The answer is A. An abnormal sweat chloride test is an expected diagnostic feature of cystic fibrosis. The chloride channel is thought to be regulated by the cystic fibrosis transmembrane regulator (CFTR) protein, which is defective in cystic fibrosis. A low C_3 complement level (choice B) may cause severe infections. The nitroblue tetrazolium (NBT) dye test (choice C) is an *in vitro* test for a respiratory burst in neutrophils. Allergic type I hypersensitivity (choice D) conditions are characterized by an increase immunoglobulin E antibodies associated with bronchial asthma. The finding of gram-positive diplococci in the sputum (choice E) is associated with *Streptococcus pneumoniae* infection.

7. The answer is B. Cystic fibrosis is a congenital autosomal recessive disease caused by multiple mutations that result in failure of the cystic fibrosis transmembrane regulator, which regulates the chloride channel, to be inserted in the plasma membrane.

CHAPTER 2
CARDIOVASCULAR PHYSIOLOGY

I. General Principles

A. The **cardiovascular system** consists of two pumps (left and right heart ventricles) and two serial circuits (pulmonary and systemic) (Figure 2–1).

1. Cardiac output from the left side of the heart (main pump)is the **systemic blood flow.**
2. Cardiac output from the right side of the heart (booster pump) is the **pulmonary blood flow.**
3. Because the two circuits are connected in series, flow (mL/min) must be equal in both; however, transient differences do occur.

B. The **systemic circuit** begins as a large vessel, the aorta, and branches into smaller vessels until **capillaries** are reached within organs.

C. **Vascular components** include arteries, arterioles, and capillaries.

1. **Arteries** are thick-walled vessels under **high pressure** that deliver oxygenated blood to the tissues.
2. **Arterioles** are the smallest branches of arteries.
   a. They have the highest resistance in the cardiovascular system and are regulated by the autonomic nervous system.
   b. Arteriolar smooth muscle tone depends on sympathetic input, local metabolites, hormones, and other mediators.
3. **Capillaries** have the largest total cross-sectional and surface areas and are the **sites of exchange** of nutrients, water, and gases.
   a. The Law of Leplace explains why capillaries can withstand high intravascular pressures.

$$T = Pr$$

where
T = tension in the vessel wall
P = transmural pressure
R = Radius of the vessel

D. In the venous circuit, **small veins (venules)** merge to form larger veins until the largest vein, the **vena cava,** returns blood to the heart.

1. **Veins** are thin-walled vessels under **low pressure** that contain most of the blood in the cardiovascular system.
2. **Venules** are the **most permeable** components of the microcirculation.

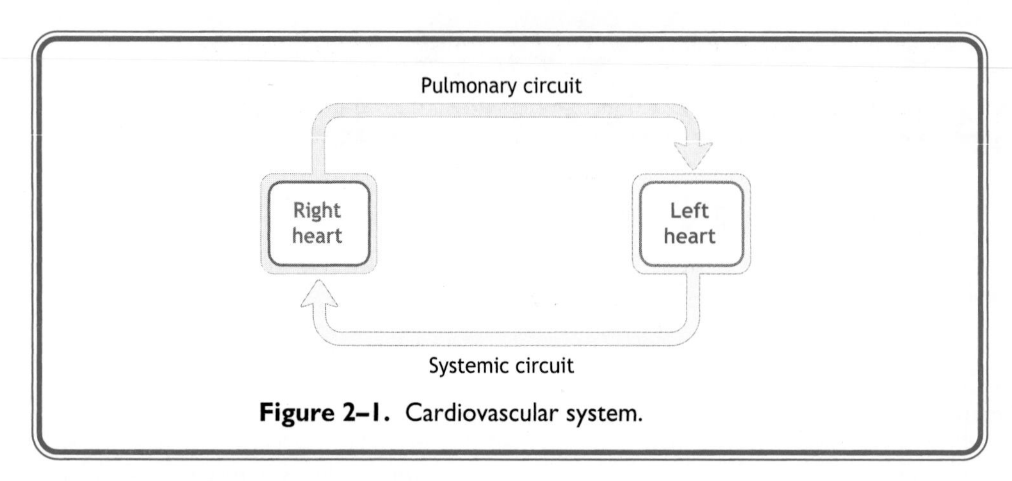

Figure 2–1. Cardiovascular system.

II. Hemodynamics

A. Velocity and Blood Flow

1. **Blood flow is driven by a constant pressure head across variable resistances.**
2. **Velocity** refers to the rate of displacement of blood within vessels with respect to time, and it has the dimensions of distance per unit time (eg, cm/s). It is expressed by the following equation:

$$V = \frac{Q}{A},$$

where
V = velocity (cm/s)
Q = blood flow (cm^3/s)
A = cross-sectional area (cm^2)

3. **Velocity** is **inversely related to** the total **cross-sectional area** of all vessels of a particular segment of the cardiovascular system.
   a. The cross-sectional area of the aorta is approximately 2.8 cm^2, whereas the area of the combined capillaries is approximately 1357 cm^2.
   b. The aorta, therefore, has the highest velocity, and the capillaries the lowest.
4. **Blood flow** is frequently designated as volume flow, and it has the dimensions of volume per unit time, for example, cubic centimeters per second.
5. Linear velocity and blood flow are then related by an area, for example, square centimeters ($cm^3/s = cm/s \times cm^2$).
6. For a given flow, the ratio of the velocity through one vessel segment relative to that in another segment depends on the inverse ratio of the respective areas:

$$\frac{V_1}{V_2} = \frac{A_2}{A_1}$$

7. This rule applies regardless of whether a given cross-sectional area pertains to a system comprising a single large tube or to a system made up of several smaller tubes in parallel.

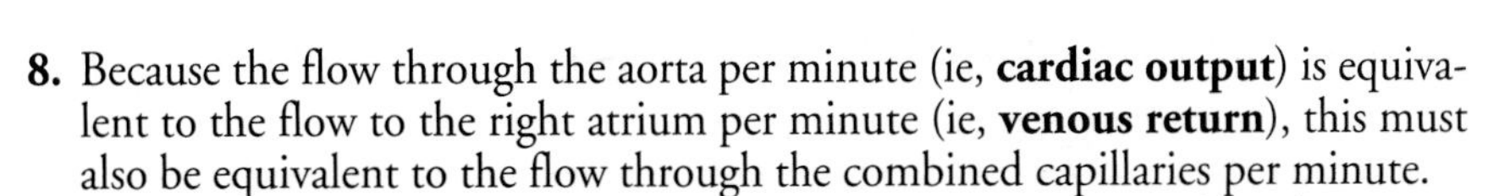

8. Because the flow through the aorta per minute (ie, **cardiac output**) is equivalent to the flow to the right atrium per minute (ie, **venous return**), this must also be equivalent to the flow through the combined capillaries per minute.
9. Total blood flow or cardiac output equals heart rate times stroke volume.

$$C.O. = F = HR \times SV$$

B. Hemodynamic Equivalent of Ohm's Law

1. The relationship between current flow and its potential difference across a conducting resistance is known as **Ohm's Law:**

$$E = IR$$

where
E = driving potential (V)
I = ionic current flow (amps)
R = resistance (ohms)

2. The equivalent relationship for a liquid in motion is

$$Q = \frac{\text{mean arterial pressure} - \text{right arterial pressure}}{\text{Total peripheral resistance}}$$

$$\Delta P = QR,$$

where
ΔP = pressure difference (mm Hg)
Q = volume flow (L/min)
R = resistance (mm Hg/L/min)

3. A driving force is required to move a flow through a resistance to flow.

C. Resistance

1. **Poiseuille's equation** gives the relationship of flow, pressure, and resistance. It considers features of the blood that are responsible for the patterns of pressure and flow through vessels:

$$Q = \frac{P_1 - P_2}{R},$$

where
Q = blood flow (L/min)
P_1 = upstream pressure for segment
P_2 = pressure at end of segment
R = resistance of vessels between P_1 and P_2

2. The equation states that flow (Q) is directly proportional to the driving pressure (ΔP) and inversely proportional to the resistance (R).
3. Resistance is directly proportional to the length (ℓ) of the vessel and to the viscosity of blood (η):

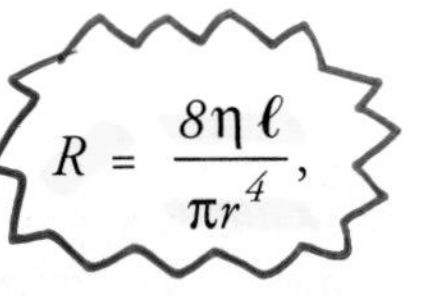

$$R = \frac{8\eta\ell}{\pi r^4},$$

where
r^4 = radius of the blood vessel to the fourth power.

a. The greater the vessel length, the greater the resistance, and the greater the viscosity, the greater the resistance.

b. The **most important factor determining resistance is the radius of the vessel.** The equation emphasizes that if the vessel radius doubles (ie, resistance decreases), then flow will increase 16-fold, if other factors remain constant.

4. The above relationship is used in conjunction with the calculation of resistance in series versus parallel circuits.
   a. To calculate **total resistance** (**R_T**) through a circulation of resistances **in series,** the individual resistances are summed (**$R_T = R_1 + R_2 + R_3$**).
   b. To calculate total resistance (**R_T**) through a circulation of resistances **in parallel,** the individual conductances are summed (**$1/R_T = 1/R_1 + 1/R_2 + 1/R_3$**).
5. Thus, if all additional parameters are held constant (eg, ΔP), a resistance change in one parallel subcircuit of the parallel circulation will not change the flow through remaining subcircuits of the parallel circulation.
6. Because the systemic and pulmonary circulations have approximately the same number of total capillaries with the same total cross-sectional area (1357 cm^2) and their blood viscosities and flows are both equal, the **lower pressure difference across the pulmonary circuit must be due to the difference in vessel length between the pulmonary and systemic circuits.**

D. Reynolds Number and Turbulence

1. **Blood flow is laminar.**
2. **Laminar flow** does not generate an audible sound; in contrast, **turbulent flow** involves random pressure fluctuations, and sounds are heard.
3. The **Reynolds number** (a dimensionless variable relating viscous and inertial forces) serves as a useful indicator for the transition of laminar flow to turbulent flow. The Reynolds number is calculated from the following equation:

$$N_{Re} = \frac{pDv}{\eta}$$

where
Re = Reynolds number
V = mean velocity (cm/s)
D = tube diameter (cm)
p = fluid density
η = fluid viscosity (Poises)

4. **Turbulent flow** primarily occurs when the Reynolds number exceeds a critical value of 3000, and flow is laminar when *Re* is below 2000.
5. Because the viscosity of blood is relatively high, the Reynolds number for turbulent flow is not exceeded in most parts of the circulation.

HEART MURMURS AND ARTERIAL BRUITS.

- *Turbulence of blood flow across diseased heart **valves create murmurs** detected by auscultation.*
- *Turbulence as blood flows through diseased arteries cause arterial murmurs known as **bruits.***

E. Compliance

1. **Compliance** describes the distensibility of blood vessels.
2. Vascular compliance (C) is the slope of the relationship between a rise in volume in the vessel and the rise in pressure produced by that rise; hence,

$$C = \frac{\Delta V}{\Delta P}$$

   a. Low compliance of a vessel leads to increased transmural pressure when the vessel blood volume is increased.
3. The compliance of combined veins is about 19 times greater than the compliance found in the combined arteries.
   a. **Systolic pressure** is a function of the stroke volume (and compliance).
   b. **Diastolic pressure** is a function of the heart rate and the arteriolar resistance, which determines run-off into the veins.

F. Pressure Profile

1. As blood flows through the systemic circulation, pressure decreases progressively from the aorta, where it is highest, to the vena cava, where it is lowest (Figure 2–2).
2. Because the greatest resistance to flow occurs in the arterioles, **the largest decrease in pressure occurs across the arterioles.**

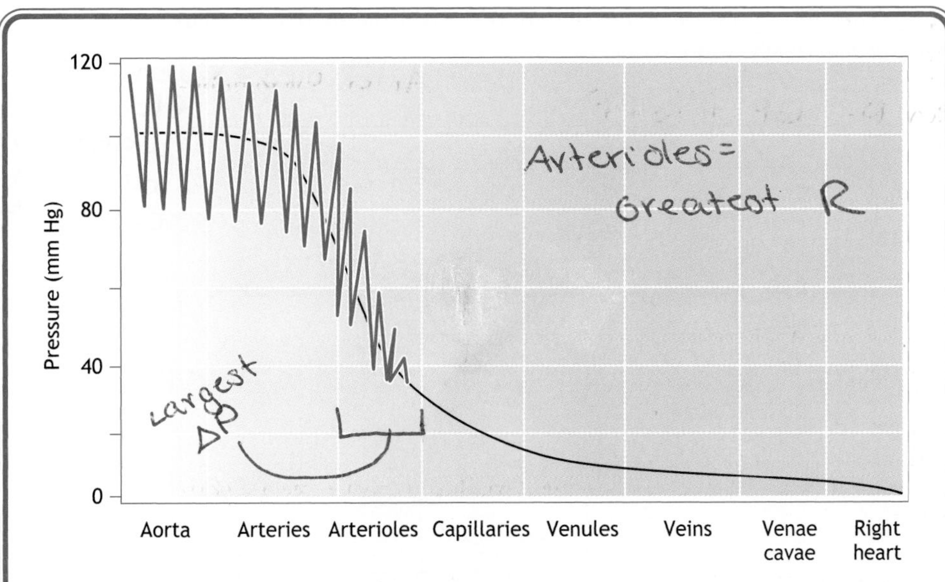

Figure 2–2. Pressure profile. Spikes in pressure represent the systolic and diastolic values during the cardiac cycle. The arterioles are the resistance vessels and dampen out the oscillations except during aortic insufficiency. Pulsatile pressures are not normally seen beyond the arteriolar level.

3. **Local arteriolar dilation** in an organ decreases arteriolar resistance, which **increases blood** flow and pressure downstream, whereas **local arteriolar constriction** increases arteriolar resistance and **decreases flow** and pressure downstream.
4. **Atrial pressure is lower than venous pressure;** pressure is 5–10 mm Hg in the left atrium and 15 mm Hg in peripheral venules.

G. Arterial Pressures (Figure 2–3)

1. **Systolic arterial pressure is the highest arterial pressure** during the cardiac cycle.
   a. It represents the pressure developed when the heart contracts most forcibly.
   b. Arterial peak systolic pressure increases, whereas minimum diastolic pressure falls as blood flows from the aorta to the peripheral arteries.
2. **Diastolic pressure is the lowest arterial pressure** during the cardiac cycle, representing the pressure when the heart is relaxed and not contracting.
3. **Pulse pressure** is the **difference between systolic and diastolic pressures** and is determined primarily by stroke volume and arterial compliance.
   a. Pressure and flow oscillate between maximum systolic and minimum diastolic valves with each heartbeat.
   b. **Pulse pressure and both arterial pressures increase with aging** due to decreased compliance of vessels.
   c. The pulse pressure also increases as blood moves out along the arterial tree.
4. **Mean arterial pressure** is the average arterial pressure over time and is calculated by adding diastolic pressure plus one third of pulse pressure.
   a. **Mean pressure,** the driving force for flow, **decreases as one moves out along the arterial tree.**
   b. The fall in mean pressure across the arteriolar bed means that capillary pressure is normally nonpulsatile.

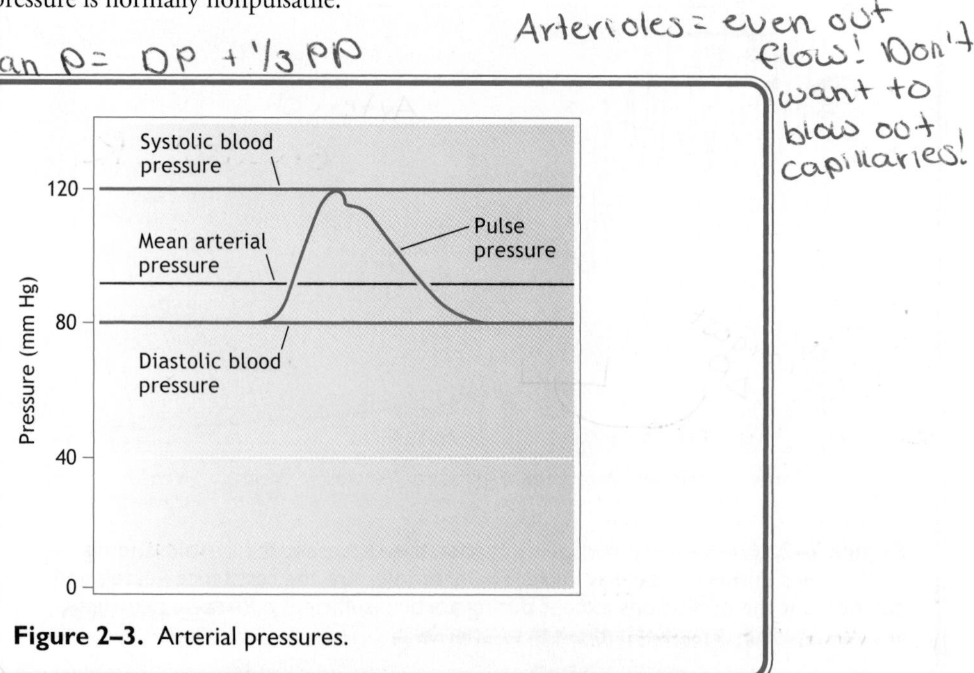

Figure 2–3. Arterial pressures.

III. Electrophysiology

A. Electrocardiogram (ECG) (Figure 2–4)

1. The **P wave** represents atrial depolarization.
2. The **PR interval** is the interval from the beginning of the P wave to the beginning of the Q wave.
   a. Prolonged PR intervals suggest a conduction delay between the atria and ventricles.
3. The **Q wave** is the beginning of ventricular depolarization.
4. The **QRS complex** represents the depolarization of the ventricles.
5. The **QT interval** is the interval from the beginning of the Q wave to the end of the T wave.
   a. A prolonged QT interval suggests drug toxicity.
6. The **ST segment** is the segment from the end of the S wave to the beginning of the T wave.
   a. An elevated ST segment suggests myocardial infarction.
7. The **T wave** represents ventricular repolarization.
   a. **An inverted T wave suggests myocardial ischemia.**

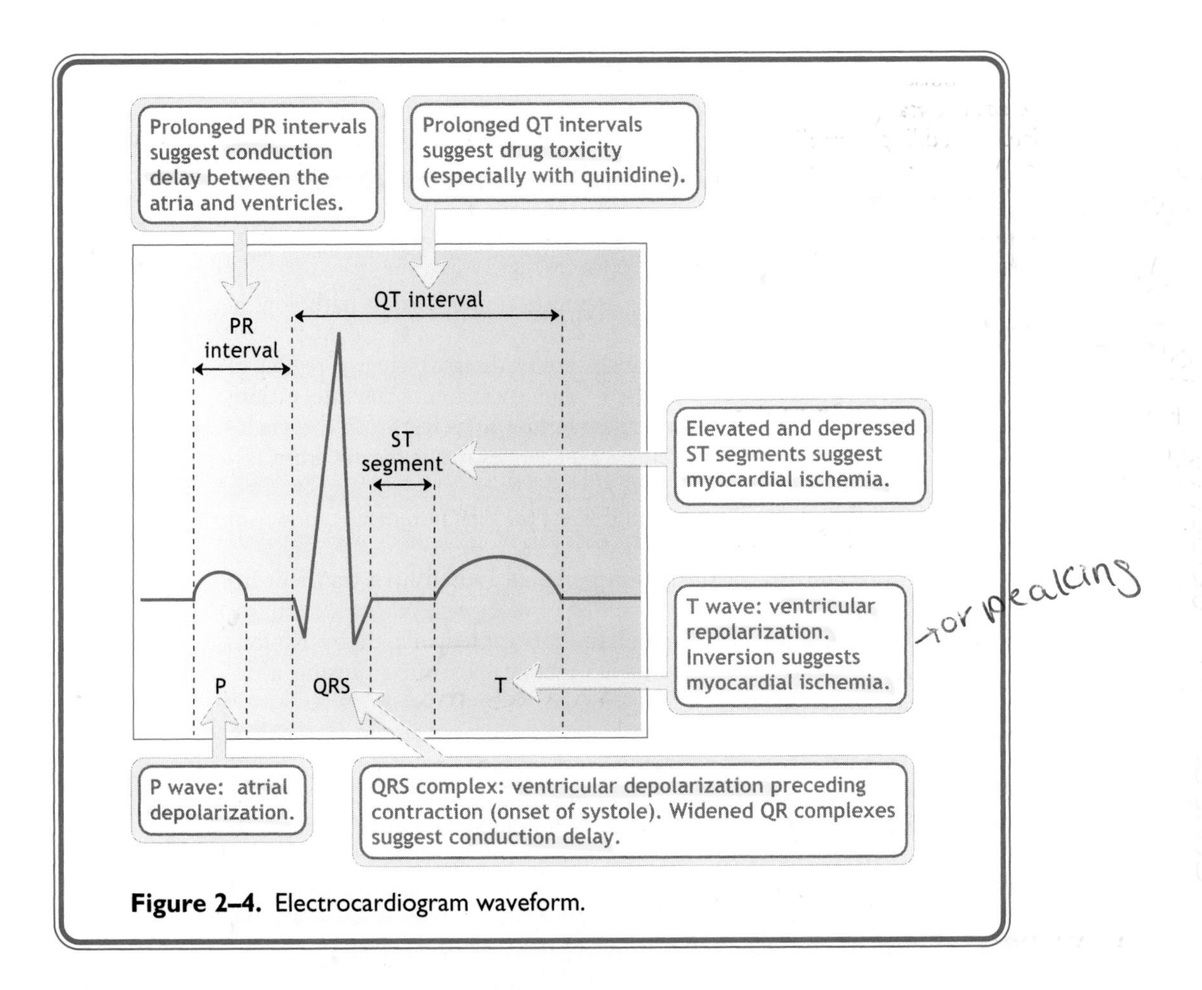

Figure 2–4. Electrocardiogram waveform.

ACUTE MYOCARDIAL INFARCTION

- *Myocardial infarction is **most commonly due to acute coronary thrombosis.***
- *It is the most common cause of death in the United States.*
- *The prognosis depends on the degree of left ventricular dysfunction.*
- *Clinical diagnosis is based on three important criteria:*
  –***Symptoms:*** *Persistent chest pain is the most common complaint. Associated symptoms include sweating, nausea, vomiting, and shortness of breath.*
  –***ECG findings:*** *Q waves (changes in ventricular depolarization), ST-segment changes (upward or downward shifts from the isoelectric line), and T-wave changes (repolarization).*
  –***Blood measurement of enzymes,*** *most commonly **creatine kinase (CK).** CK isoenzymes are composed of M and B polypeptides, and high concentrations of CK-MB indicates myocardial damage.*
- *A Q wave is the initial negative deflection in the QRS complex, and a large Q wave is diagnostic of a myocardial infarction.*
- *The ST segment correlates with phase 2, or the plateau phase, of ventricular myocytes, and myocardial infarction leads to persistent ST-segment elevation when the positive electrode lies over the injured area.*
- *Myocardial ischemia can be associated with repolarization abnormalities reflected in T-wave inversion or T-wave peaking (spikelike).*
- ***Initial treatment*** *involves morphine for pain, thrombolytic therapy within 6 hours, heparin anticoagulation and intravenous nitrates, and β-blockers to decrease acute morbidity.*

B. Action Potentials

1. Myocardial cell **action potentials** are classified, according to their shapes, as fast or slow responses.
   a. The **fast response** occurs in ordinary **atrial and ventricular myocytes** and in specialized conducting fibers (**Purkinje fibers**).
   b. The **slow response** is found in the **sinoatrial** (**SA**) **node** and the **atrioventricular** (**AV**) **node.**
   c. Fast responses may change to slow responses under certain pathological conditions.
   d. For example, in patients with coronary artery disease, when a region of cardiac muscle becomes ischemic, the K^+ concentration in the interstitium that surrounds the affected muscle rises because K^+ is lost from the inadequately perfused (ie, ischemic) cells. This changes the myocytes from fast to slow responders.
2. Action potentials of ventricles, atria, and the Purkinje system are shown in Figure 2–5.
   a. **Phase 0** is the upstroke of the action potential caused by a transient increase in Na^+ conductance due to opening of voltage-gated Na^+ channels.
   b. **Phase 1** is a period of partial repolarization caused in part by K^+ ions moving out of the cell. Na^+ channels become inactivated causing a decrease in Na^+ conductance.
   c. **Phase 2** is the plateau phase caused by a transient increase in Ca^{2+} influx through the voltage-gated Ca^{2+} channels. K^+ efflux remains low.
   d. **Phase 3** is a period of rapid repolarization caused partly by a large K^+ outward current, which hyperpolarizes the membrane, and partly by inactivation of the Ca^{2+} channels.
   e. **Phase 4** is the resting membrane potential; the membrane potential is near the K^+ equilibrium potential. It represents the interval from the end of repolarization until the beginning of the next action potential.

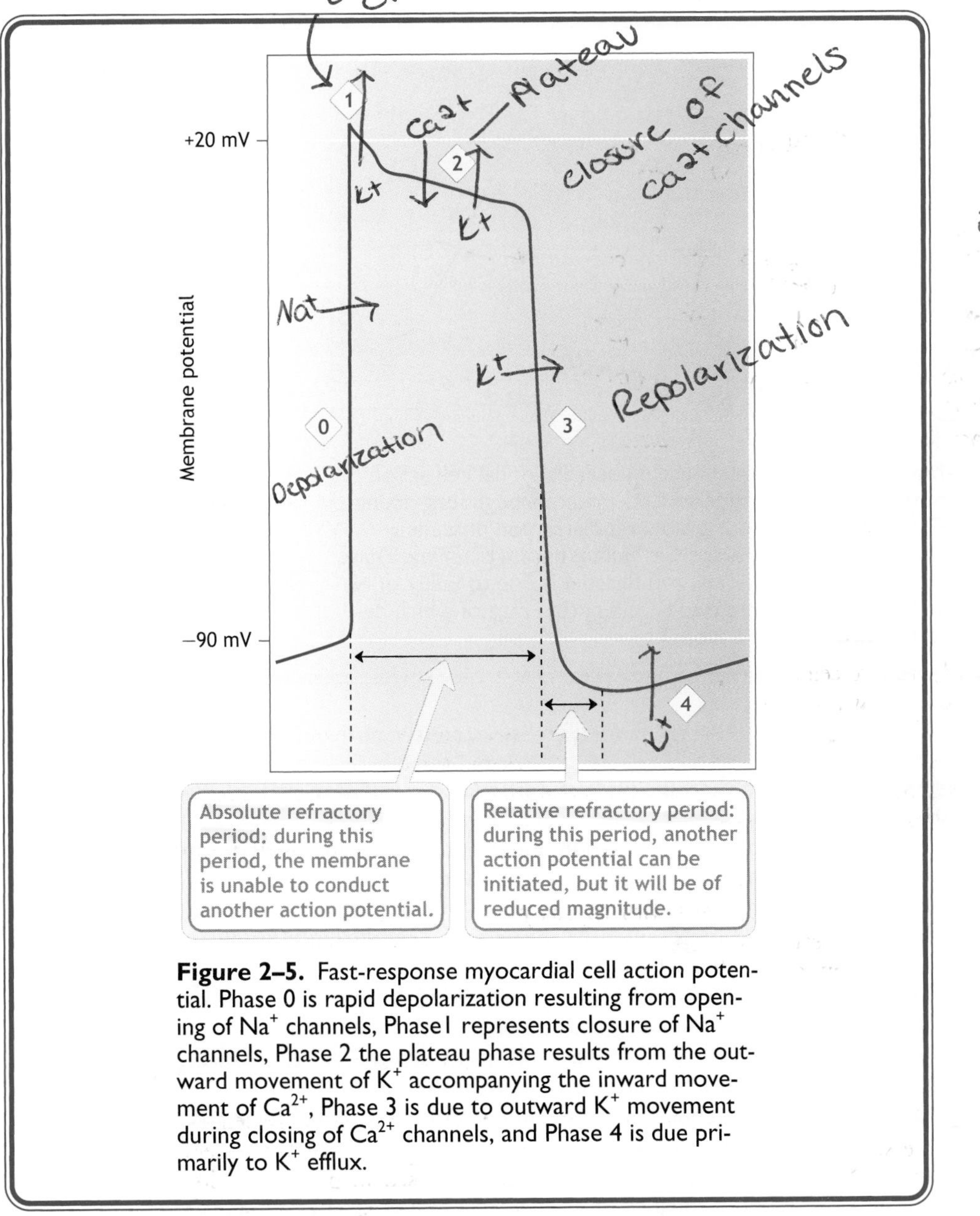

Figure 2–5. Fast-response myocardial cell action potential. Phase 0 is rapid depolarization resulting from opening of Na^+ channels, Phase I represents closure of Na^+ channels, Phase 2 the plateau phase results from the outward movement of K^+ accompanying the inward movement of Ca^{2+}, Phase 3 is due to outward K^+ movement during closing of Ca^{2+} channels, and Phase 4 is due primarily to K^+ efflux.

3. Extracellular calcium influences the action potential plateau.

a. **Phase 2,** the plateau phase, is achieved by a balance between the influx of Ca^{2+} through Ca^{2+} channels and the efflux of K^+ through several types of K^+ channels.

b. **Phase 3,** final repolarization, is initiated when the efflux of K^+ exceeds the influx of Ca^{2+}. Hence, calcium channel antagonists (eg nifedipine) decrease the amplitude and duration of action potentials.

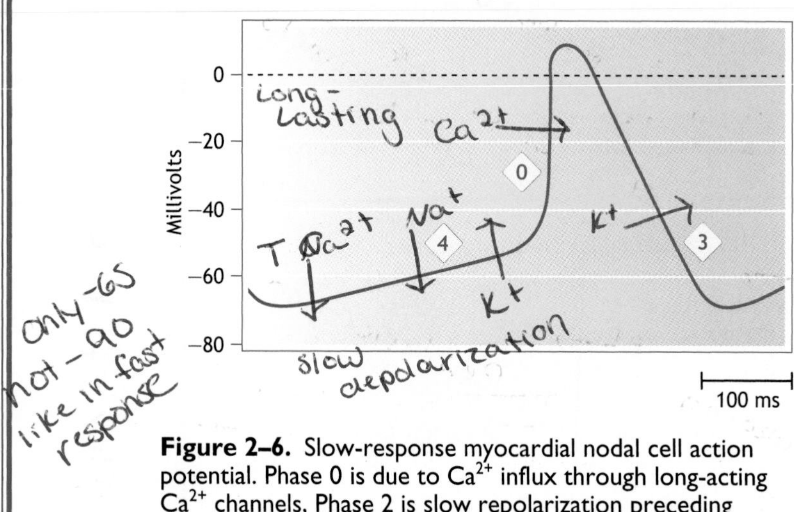

Figure 2–6. Slow-response myocardial nodal cell action potential. Phase 0 is due to Ca^{2+} influx through long-acting Ca^{2+} channels, Phase 2 is slow repolarization preceding Phase 3 due to decreased Ca^{2+} influx during K^+ efflux, Phase 3 is due to K^+ efflux only, and Phase 4 is due to influx of Na^+ and Ca^{2+} during decreased K^+ efflux (the rate of which determines prepotential slope).

4. The sinoatrial (**SA**) **node** is the primary pacemaker of the heart and exhibits phase 4 depolarization, which is responsible for its automaticity (Figure 2–6).
   a. **Phase 0** is the upstroke of the action potential caused by increased inward Ca^{2+} current. This phase also occurs in the **AV node.**
   b. **Phases 1** and **2** are not present.
   c. **Phase 3** is repolarization caused by increased K^+ conductance producing an outward K^+ current.
   d. **Phase 4** is slow depolarization, which accounts for the pacemaker activity of the SA node and is caused by an increase in Na^+ conductance.
      (1) The increased Na^+ conductance results in an **inward Na^+ current, I_f.**
      ***(2)* The Na+ current is the largest current in the heart.**
      (3) **I_f** is initiated by repolarization of the membrane potential during the previous action potential.
   e. The **transmembrane potential** (maximum diastolic potential, MDP) during phase 4 of SA nodal cells is much less negative because the I_K type of K^+ channel is sparse in these cells.
   f. The conduction velocities of the slow responses of the SA and AV nodes are about 0.02–0.1 m/s. These cells are designed to initiate an action potential.
   g. The fast-response conduction velocities are about 0.3–1.0 m/s for atrial and ventricular contractile cells, and 1.0–4.0 m/s for specialized conducting fibers in the atria and ventricles.
   h. Regions of the heart other than the SA node may initiate beats under special circumstances: such sites are called **ectopic foci,** or **ectopic pacemakers.** Ectopic foci become pacemakers when

(1) Their own rhythmicity becomes enhanced

(2) The rhythmicity of the higher-order pacemakers becomes depressed (The AV node and Purkinje systems may replace the SA node if it is suppressed.)

(3) All conduction pathways are blocked between the ectopic focus and those regions with greater rhythmicity

C. Cardiac Action Potential Fluxes of Sodium, Potassium, and Calcium

1. The Na^+ current (I_{NA}) is responsible for the rapid depolarizing phase of the action potentials in atrial and ventricular muscle and Purkinje fibers.

2. In the slow-response nodal cells, the diastolic depolarization is mediated by at least three ionic currents:

a. An **inward pacemaker current, I_f,** is induced by hyperpolarization and carried mainly by Na^+ and Ca^{2+}. This current is mediated by a nonselective cation channel that differs from the fast-response Na^+ channels.

b. A **calcium channel** and **the calcium current, I_{Ca},** become activated toward the end of phase 4. The Ca^{2+} current passes through L-type Ca^{2+} channels in the heart.

(1) The **influx of Ca^+** accelerates the rate of diastolic depolarization, leading to the upstroke of the action potential through separate Ca^{2+} membrane channels.

(2) A **decrease in extracellular Ca^{2+}** concentration or administration of a calcium channel antagonist diminishes the amplitude of the action potential and the slope of the pacemaker prepotential (which precedes the threshold potential).

c. An **outward K^+ current, I_K,** tends to repolarize the cell after the upstroke of the action potential. The autonomic neurotransmitters alter the ionic currents across the cell membranes. This repolarizing K^+ current turns on slowly.

(1) The **adrenergic-mediated** decrease in MDP and increase in depolarization indicates that the **increases of I_f and I_{Ca}** must exceed the enhancement of I_K.

(2) **Acetylcholine depresses I_f** and I_{Ca} and increases the MDP.

(3) L-type Ca^{2+} channel blocking drugs act by inhibiting I_{Ca}.

D. Dual Innervation of the Heart

1. The frequency of pacemaker firing is controlled by the activity of both divisions of the autonomic nervous system (**dual innervation**).

a. **Increased sympathetic nervous activity,** through the release of norepinephrine, **raises the heart rate** principally by **decreasing the rate of K^+** efflux during the diastolic depolarization.

b. **Increased vagal activity,** through the release of acetylcholine, **decreases the heart rate** by **increasing the rate of K^+** efflux during the diastolic depolarization.

2. Over an intermediate range of arterial pressures (approximately 10–200 mm Hg), **the alterations in heart rate are achieved by reciprocal changes in vagal and sympathetic neural activity to the SA and AV nodes.**

3. Below the 10–20 mm Hg range of arterial blood pressures, high heart rate is achieved by **intense sympathetic activity** and the virtual absence of vagal activity.

4. Above the 200 mm Hg range of arterial pressures, low heart rate is achieved by **intense vagal activity** and a low level of sympathetic activity.

5. When the vagus nerve is severed, the central end is connected to the medulla oblongata, and the peripheral end innervates the myocardium.
   a. Stimulation of the peripheral end produces a low heart rate similar to a major vagal stimulation of the heart.
   b. Depending on the frequency of stimulation, one can merely slow the heart or produce a complete AV block.
   c. An **AV block** will result in an ectopic ventricular pacemaker eventually taking over, which is known as **vagal escape.** Thus, the ventricular tissue escapes the influence of intense vagal stimulation.

E. Refractory Time for Cardiac Muscle Fiber Types

1. The action potential recorded from **Purkinje fibers exhibits a long plateau period.**
2. Because of the long refractory period of the Purkinje fibers, many premature activations of the atria are conducted through the AV junction but are blocked by the Purkinje system.
3. As heart rate increases, the refractory period diminishes.

IV. Cardiac Muscle and Cardiac Output

A. Ventricular Action Potential versus Mechanical Events

1. The **QRS complex** is the body surface simultaneous recording of all ventricular cell **phase 1 depolarizations.**
2. The **T wave** is the body surface simultaneous recording of all ventricular cell **phase 3 repolarizations.**
3. The **T wave** begins midway through the ejection phase and **continues until the onset of the isovolumetric relaxation phase.**

B. Myocardial Cell Structure

1. Cardiac muscle cells contain numerous **myofibrils,** which are chains of **sarcomeres,** the fundamental contractile unit.
2. Myocytes are coupled to one another by **intercalated disks.**
3. Although the myocardium is made up of individual cells with discrete membrane boundaries, the cardiac myocytes that comprise the ventricles contract almost in unison, as do those in the atria.
4. **Cell-to-cell conduction** occurs through **gap junctions,** which are **low-resistance pathways** that are a part of the intercalated discs and allow for rapid electrical spread of action potentials to cells.
5. Cardiac muscle differs from skeletal muscle in the following ways:
   a. Cardiac muscle contains only one or two centrally located nuclei, in contrast to the several nuclei in skeletal muscle.
   b. Gap junctions are found only in cardiac muscle.
   c. Compared to skeletal muscle, cardiac muscle contains fewer but larger T-tubules, particularly in the atria.

C. Similar Cardiac Output: Right and Left Heart

1. The **stroke volume** (SV) of the two ventricles must, at steady state, be **identical.**
2. The **rate** (HR) of the two ventricles must be identical.
3. Hence, the **output** (HR × SV) of the two ventricles must also be identical.
4. In the steady state, output of the left ventricle is recorded over a 1-minute interval and is termed the **cardiac output,** or CO:

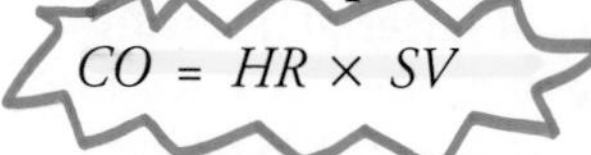

$$CO = HR \times SV$$

5. **Cardiac output is equivalent to the venous return.** For example, cardiac output increases during exercise because of the fall in skeletal muscle resistance and increased venous return.

D. Excitation-Contraction Coupling

1. This coupling links the electrical activities of the myocyte to the force-generating actin-myosin reaction.
2. Ca^{2+} enters the myocyte mainly during phase 2 of an action potential via voltage-activated channels.
3. This Ca^{2+} entry triggers the release of Ca^{2+} from intracellular **sarcoplasmic reticulum (SR)** stores, **increasing intracellular Ca^{2+}** levels.
4. Ca^{2+} binds to troponin C, moving tropomyosin away and allowing actin and myosin binding.
5. Actin and myosin bind, the thick and thin filaments slide past one another, and the myocardial cell contracts.
6. The **strength of contraction** correlates with the amount of **SR Ca^{2+}** release.
7. **Ca^{2+} removal** by an active Ca^{2+}-ATPase pump **is required for relaxation.**

E. End-Diastolic Blood Pressure Changes with Change in Cardiac Output

1. **Changes in cardiac output** are generally **brought about by changes in autonomic activity.** Hence, with an **increase in sympathetic activity,** the **rate** and the myocardial contractility (ie, **stroke volume**) will **increase.**
2. The result will be **decreased ventricular end-diastolic pressure,** because these induced cardiac changes are accompanied by a concurrent increase in arteriolar resistance (ie, **vasoconstriction**).
3. With the increase in cardiac output during exercise, ventricular end-diastolic pressure will not decrease, as a result of reduction in peripheral resistance from dilation in the skeletal muscle beds.

F. Starling's Law (Figure 2–7)

1. The relation between fiber length and strength of contraction is known as **Starling's law of the heart.**
2. An increase in myocardial fiber length, as occurs with an increased ventricular filling during **diastole** (ie, **preload**), produces a more forceful ventricular contraction because more overlap between thick and thin filaments is exposed for cross-bridge formation.
3. Hence, a **decreased heart rate,** with longer filling time, will result in an **increase in stroke volume.**
4. Starling's law is active only to the point at which a maximal systolic pressure is reached at the optimal preload.
5. If diastolic pressure increases beyond the optimal preload, no further increases in developed pressure will occur. Thus, the normal heart operates on the ascending portion of the Frank-Starling curve.
6. Certain pharmacologic agents increase contractility and cardiac output (ie, positive inotropic agents such as digitalis or catecholamines), whereas others decrease contractility and cardiac output (ie, negative inotropic agents such as β-blockers (propanolol), or calcium channel blockers (nifedipine).

G. Pressure-Volume Loop of the Left Ventricle

1. **The opening and closing of the cardiac valves define the four phases of the cardiac cycle**
   a. **Isovolumetric contractions**

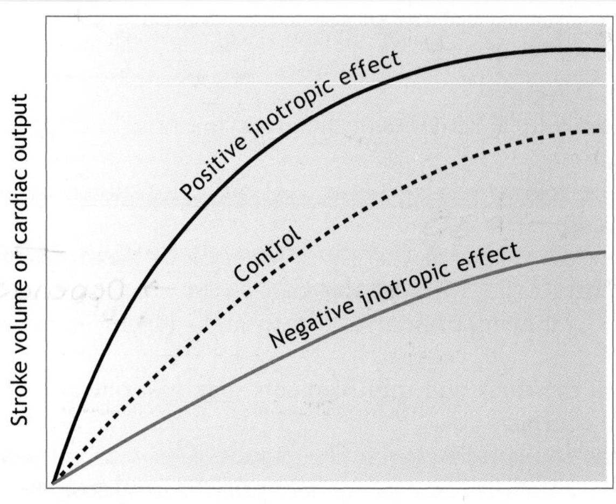

Figure 2–7. Starling's law. A positive inotropic effect produces an increase in peak force developed during contraction. A negative inotropic effect is the opposite (due to changing Ca^{2+} concentrations during contraction). The three lines represent Starling curves with changes in force developed and their influence on stroke volume at any given preload (end diastolic volume).

b. **Outflow phase (ventricular ejection)**
c. **Isovolumetric relaxation**
d. **Inflow phase (ventricular filling)**

W=PV

2. The **external work of the heart** can be approximated as the product of pressure (P) times stroke volume (SV), or more accurately as the integral $\int PdV$, which is the pressure-volume loop of the heart.
3. A single left ventricular cycle of contraction, ejection, relaxation, and refilling is visualized in the **pressure-volume loop** (Figure 2–8):
   a. **Isovolumetric contraction** is represented in the figure by movement from points C to D.
      (1) Point C is diastole with the ventricular muscle relaxed and filled with blood to about 145 mL (end-diastolic volume).
      (2) Upon excitation, the ventricle contracts but no blood is ejected because all of the valves are closed.
   b. **Ventricular ejection** (outflow phase) is represented by movement from point D to F.
      (1) At point D the aortic valve opens and blood is ejected into the aorta.
      (2) The volume ejected per beat is the stroke volume and is graphically depicted by the width of the pressure-volume loop.
      (3) Point F is the end-systolic volume.
   c. **Isovolumetric relaxation** is represented by movement from points F to A.

A STANDARD CONTRACTILITY CONDITIONS

Theoretical maximal isovolumetric pressure

Left ventricular pressure

ESPVR

Left ventricular volume

B INCREASED CONTRACTILITY

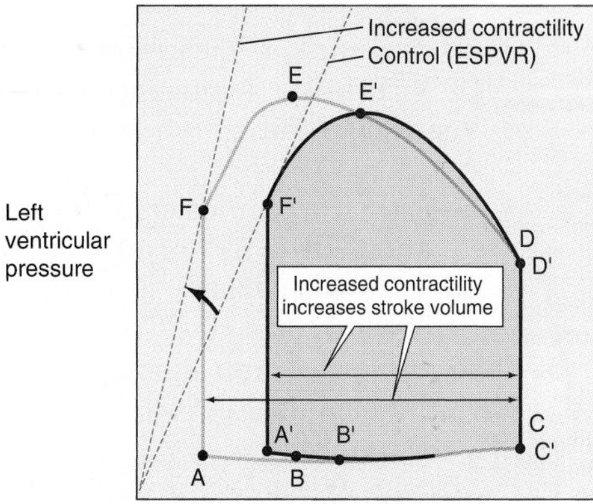

C INCREASED PRELOAD (FILLING)

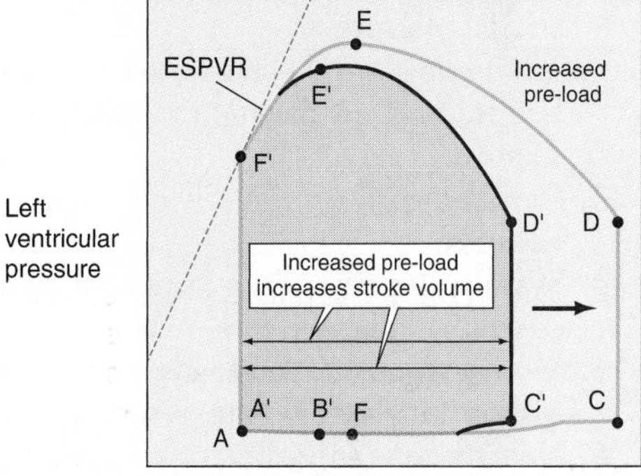

D INCREASED AFTER-LOAD (AORTIC PRESSURE)

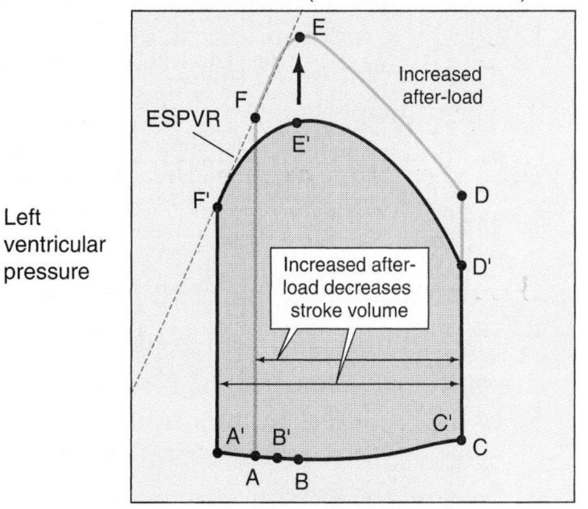

Figure 2–8. Assessment of contractility using a ventricular pressure-volume loop. The A′,B′,C′,D′,E′,F′ pressure-volume loop is the normal curve. In A, at the same normal state of cardiac contractility, the A,B,C,D,E,F pressure-volume loop is generated by decreasing end-diastolic volume (EDV) and the A″,B″,C″,D″,E″,F″ loop is generated by increasing EDV. The slope of the line through the points at the end of systole (F, F′, and F″) represents the end-systolic pressure-volume relation (ESPVR). *(Reprinted from Boron and Boulpaep: Medical Physiology, Figure 21–11, Page 531 © 2003 with permission from Elsevier, Inc.)*

(1) At point F, as the ventricle relaxes, the aortic valve closes.
(2) Ventricular volume is constant because all valves are closed.

d. **Ventricular filling** (inflow phase) is represented by movement from point A back to C.
(1) After left ventricular pressure decreases below left atrial pressure, the mitral valve (AV) opens and filling begins.
(2) Ventricular volume increases to about 140 mL (end-diastolic volume), of which only 10–20% results from atrial contraction.

3. Abnormalities of both filling and emptying usually coexist when the heart fails; these abnormalities can be seen in the pressure-volume loop.

H. Cardiac Work (see Figure 2–8)

1. **Cardiac work** is the amount of work done by the heart on each beat.
2. Even though the output of the right and left heart is equal, cardiac work is much greater for the left heart because of the greater **afterload,** or increase in arterial pressure.
   a. Afterload on the left ventricle is equivalent to **aortic pressure.**
   b. Afterload on the right ventricle is equivalent to **pulmonary artery pressure.**
3. **Cardiac work** is primarily a function of arterial **systolic pressure** and **stroke volume.**
4. **Systolic pressure is a function of stroke volume.** As stroke volume increases, systolic pressure increases.
5. With **increased afterload,** the ventricle must work harder to eject blood against a higher pressure, resulting in a **decrease in stroke volume.**
6. **Heart rate** is an **indicator of stroke volume,** because as heart rate increases, stroke volume usually decreases, due to decreased filling time.

I. Fick Principle

1. The **Fick method** for calculating cardiac output is an application of the **law of conservation of mass.**
2. The principle states that the O_2 delivered to the pulmonary capillaries via the pulmonary artery, plus the O_2 that enters the pulmonary capillaries from the alveoli, must equal the quantity of O_2 carried away by the pulmonary veins.
3. The cardiac output is calculated by dividing the pulmonary O_2 uptake per minute by the difference between systemic arterial O_2 content (mL O_2/100 mL blood) and pulmonary arterial O_2 content (mL O_2/100 mL blood).
4. The denominator represents the pulmonary arteriovenous O_2 difference in volumes percent (mL O_2/100 mL blood, or vol%).
5. In the clinical determination of cardiac output, O_2 consumption is computed by measuring the volume of O_2 content expired over a period of time.
6. Because the O_2 concentration of peripheral arterial blood is essentially identical to that in the pulmonary veins, arterial O_2 concentration is determined by using a sample of peripheral arterial blood.

J. Venous Return and Central Venous Pressure

1. The **venous return** (ie, vascular function) relationship defines the changes in central venous pressure evoked by changes in cardiac output.
   a. Increased venous return stretches low pressure atrial receptors causing tachycardia which is known as the **Bainbridge Reflex.**

2. As **cardiac output increases,** blood is removed from the central veins at a greater rate, and **central venous pressure and right atrial pressure decreases.**
3. **Central venous pressure** is the **response,** and **cardiac output** is the **stimulus.**
4. This relationship contrasts with the cardiac function relationship using the Frank-Starling mechanism, in which **central venous pressure** (ie, preload) **is the stimulus,** which drives a larger **cardiac output** (ie, **the response**) by its influence on stroke volume.
5. When the system is in equilibrium, cardiac output is equivalent to venous return.

V. Cardiac Cycle with Pressures and ECG

A. Simultaneous recording of left atrial, left ventricular, and aortic pressures; heart sounds; ventricular volume; venous pulse; and the ECG graphically portray the sequential and related electrical and cardiodynamic events throughout a **cardiac cycle** (Figure 2–9). One must be able to reproduce this figure from memory.
   1. The **P wave** on the ECG precedes **atrial systole,** which contributes to ventricular filling that causes the **fourth heart sound.**
   2. **Atrial systole** increases venous pressure, which is represented by the **a wave** on the venous pressure curve.
   3. **Isovolumetric ventricular contraction** begins after the onset of the QRS complex, leading to increased ventricular pressure and **closure of the AV valves,** which corresponds to the **first heart sound.**
   4. **Rapid ventricular ejection** occurs when ventricular pressure reaches a maximum and the aortic valve opens, releasing most of the stroke volume.
   5. **Isovolumetric ventricular relaxation** occurs with repolarization of the ventricles and **closure of the aortic valve** and **pulmonic valve,** which corresponds to the **second heart sound.**
      a. The indentation on the aortic pressure tracing following closure of the aortic valve is called the **incisura.**
      b. When ventricular pressure becomes less than atrial pressure the **mitral valve** opens.
   6. **Rapid ventricular filling** occurs after the mitral valve opens, and the rapid flow from the atria into the ventricles causes the **third heart sound.**

B. The electrical and cardiodynamic events portrayed in Figure 2–9 help explain why **mitral insufficiency and narrowing** (ie, **stenosis**) produce systolic and diastolic murmurs, respectively.

C. **Aortic insufficiency and stenosis** produce diastolic and systolic murmurs, respectively.

D. These murmurs are heard best in the 2nd intercostal space just to the right of the sternum.

AORTIC REGURGITATION

- ***Rheumatic fever*** *is the most common cause of aortic regurgitation.*
- *Volume overload in the left ventricle, due to ischemia or valvular problems, leads to left ventricular dilatation and hypertrophy.*

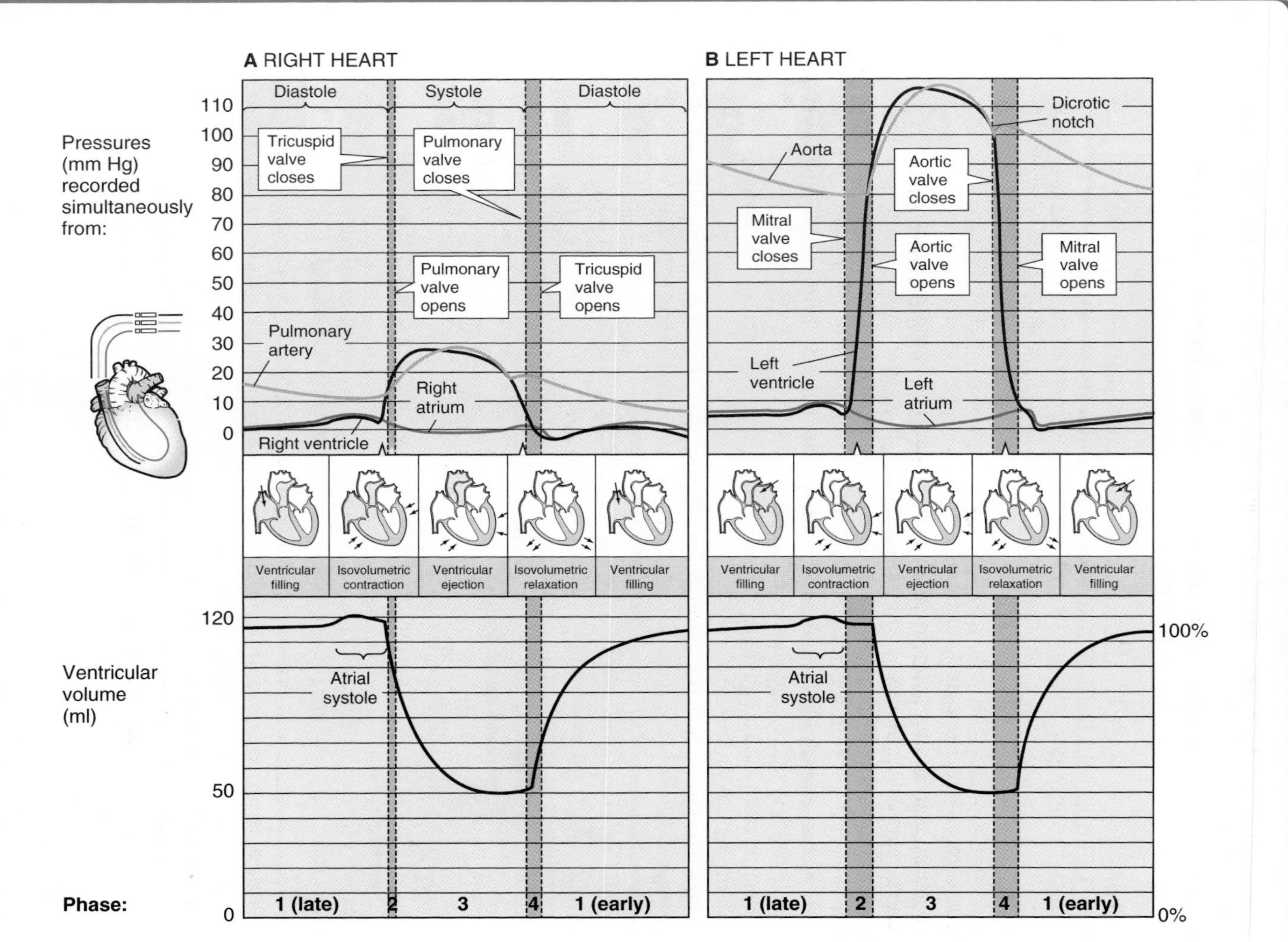

Figure 2–9. Pressures and ventricular volumes during the cardiac cycle. A. Right heart. B, Left heart. The insert shows the placement of catheters used for pressure measurements in the right heart. *(Reprinted from Boron and Boulpaep: Medical Physiology, Figure 21–1, Page 510 © 2003 with permission from Elsevier, Inc.)*

- *Stroke volume and pulse pressure are increased.*
- *A clinical diagnostic feature is an **early diagnostic murmur** along the 2nd and 3rd intercostal spaces, produced by regurgitation of blood into the left ventricle.*
- ***Widened pulse pressure** occurs due to the combination of a drop in diastolic blood pressure as blood flows back into the left ventricle and an increase in systolic pressure from increased stroke volume.*

VI. Regulation of Arterial Pressure

A. Baroreceptors versus Chemoreceptors

1. The **baroreceptors** (pressoreceptors) in the walls of the carotid sinus near the internal carotid arteries and in the aortic arch are tonically active and regulate blood pressure on a moment-to-moment basis.
2. Stretching these receptors by **increased arterial pressure** reflexively **induces bradycardia** and **vasodilation.**
3. A **decrease in arterial pressure relieving the stretch on high-pressure arterial receptors causes tachycardia** and **vasoconstriction.**
4. Baroreceptors are also present in the cardiac chambers and large pulmonary vessels (**cardiopulmonary baroreceptors**), where they participate in blood volume regulation.
5. Stimulation of **peripheral chemoreceptors** found at the **carotid and aortic bodies** and of **central chemoreceptors** found in the medulla oblongata—via a decrease in blood O_2 tension and an increase in blood CO_2 tension—increases the rate and depth of respiration and produces peripheral vasoconstriction.
6. The directional change in heart rate evoked by peripheral chemoreceptor stimulation is proportional to the change in respiratory rate. **As respiratory minute volume is increased, heart rate is increased** and vice versa.

B. Renin-Angiotensin System (Figure 2–10)

1. The **renin-angiotensin system** is a hormonal mechanism for long-term blood pressure regulation through adjustment of blood volume.
2. **Renin** secretion from **juxtaglomerular cells** of the renal afferent arteriole is increased by
   a. Decreased stretch of the afferent arteriolar wall
   b. Decreased distal tubular delivery of NaCl to the macula densa cells
   c. β_1-Adrenoreceptor activation by sympathetic nerves supplying the juxtaglomerular apparatus
3. The enzyme **renin** converts circulating **angiotensinogen** to **angiotensin I.**
4. **Angiotensin I,** which has **no biologic activity,** is converted primarily in endothelial cells of the lungs by **angiotensin-converting enzyme** (ACE) to **angiotensin II.**
5. **Angiotensin II** is a potent **vasoconstrictor** and a promoter of **aldosterone** secretion from the **adrenal zona glomerulosa.**
6. The vasoconstriction of the arterioles increases **total peripheral resistance** and **mean arterial pressure.**
7. Angiotensin II is converted by an **aminopeptidase** to **angiotensin III,** which also acts on the zona glomerulosa of the adrenal gland to promote **aldosterone** secretion.
8. **Aldosterone** increases NaCl reabsorption by the renal distal tubule, thereby **increasing blood volume** and **arterial pressure.**

$Q = P/R$

Water follows Na^+ Cl^-

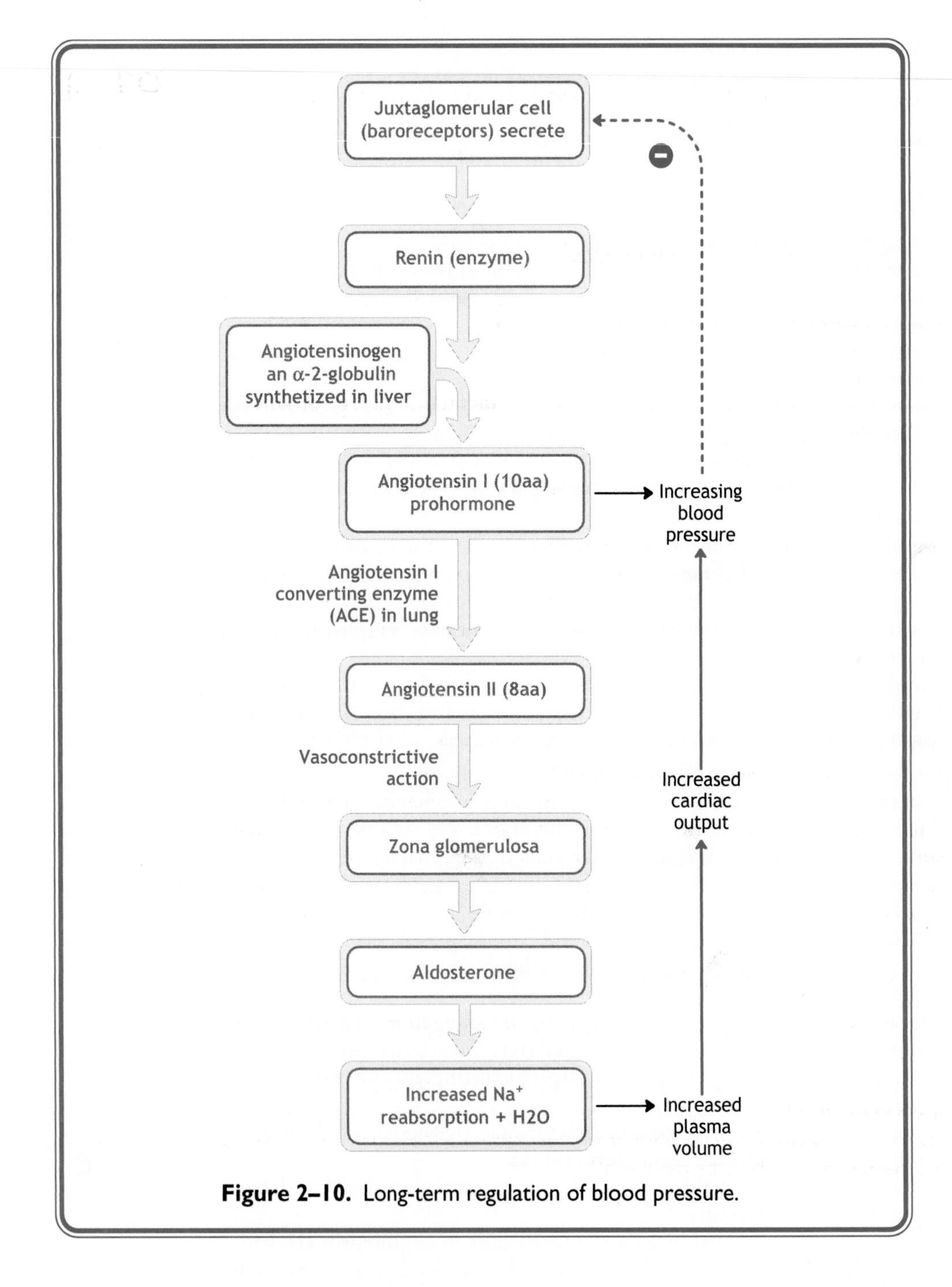

Figure 2–10. Long-term regulation of blood pressure.

HYPERTENSION

- *Hypertension is defined as a chronic elevation of arterial blood pressure and is the result of dysfunction of mechanisms used for long-term control of arterial pressure.*
- *Ninety percent of cases are essential (primary) hypertension where no specific cause has been identified.*
- *Chronic hypertensives may have diminished sensitivity of their arterial baroreceptors.*
- *The most common cause of secondary hypertension is renal artery stenosis.*

VII. Control Mechanisms and Special Circulations

A. Autoregulation

1. **Autoregulation** is the maintenance of constant blood flow over a wide range of blood pressures.
2. Constant flow is due to
   a. Increases or decreases in local metabolites (**metabolic theory of autoregulation**)
   b. Smooth muscle contraction in response to increases or decreases in pressure (**myogenic theory of autoregulation**)

B. Active Hyperemia

1. **Active hyperemia** is defined as increased blood flow to an organ caused by increased tissue metabolic activity and accumulation of vasodilator metabolites.
2. In exercise, blood flow will increase to skeletal muscles involved to meet increased metabolic demand.

C. Reactive Hyperemia

1. If arterial inflow to a vascular bed is stopped for a few minutes, the blood flow, on release of the occlusion, immediately exceeds the flow before the occlusion, producing **reactive hyperemia.**
2. A number of metabolites may mediate the metabolic vasodilation that occurs during the interval of occlusion, such as **CO_2**, H^+, K^+, lactic acid, and **adenosine.** The increase in flow is proportional to the length of the occlusion. The relative contribution of these metabolites remains the subject of future investigation.

D. Coronary Blood Flow

1. The **principal factor** responsible for perfusion of the myocardium is **aortic pressure.**
   a. **Extravascular compression impairs coronary blood flow during systole**
2. Changes in coronary blood flow are caused mainly by caliber changes of the **coronary resistance vessels** in response to metabolic demands of the heart.
3. **A decrease in O_2** supply or an increase in O_2 demand apparently **causes the release of a vasodilator** (**adenosine**) that decreases coronary resistance and increases coronary flow proportionally.

E. Cutaneous Circulation and Temperature

1. The skin contains two types of resistance vessels: **arterioles** and **arteriovenous anastomoses.**
2. The **arterioles** are similar to those found elsewhere in the body.
3. **Arteriovenous anastomoses shunt blood** from the **arterioles to venules** and venous plexuses, bypassing the capillary bed.
   a. **Arteriovenous anastomoses** are found primarily in fingertips, palms of the hand, soles of the feet, ears, nose, and lips (ie, exposed regions).

b. These vessels are almost exclusively under **sympathetic neural control by temperature receptors** from higher centers and become maximally dilated when their nerve supply is interrupted.
c. They do not appear to be under metabolic control, and they **fail to exhibit reactive hyperemia or autoregulation.**

F. Fetal Circulation at Birth (Figure 2–11)

1. In the fetus, blood returning to the right heart is divided into two streams by the edge of the interatrial septum (**crista dividens**).
2. The **larger stream** is **shunted to** the **left atrium** through the **foramen ovale.**
3. The other stream passes into the **right atrium,** where it is joined by superior vena cava blood returning from the upper parts of the body.
   a. The fetal circulation has four shunts: the placenta, ductus venosus, foramen ovale, and ductus arteriosus.
4. Because of the large pulmonary resistance, due to the low fetal partial pressure of O_2 in alveolar gas, only **one-tenth of the right ventricular output goes through the lungs.**
5. The remainder passes through the **ductus arteriosus** from the pulmonary artery to the descending aorta. Blood flows from the pulmonary artery to the aorta because the pulmonary resistance is high and the diameter of the ductus arteriosus is as large as the descending aorta.
6. **At birth,** the asphyxia that starts with clamping of the umbilical vessels activates the infant's respiratory center.
7. As the lungs fill with air, pulmonary vascular resistance decreases to about one-tenth of the value existing before lung expansion.
8. The **left atrial pressure** is raised above the pressure in the inferior vena cava and right atrium, and this reversal of the pressure gradient across the atria abruptly closes the valve over the **foramen ovale.**
9. With the decrease in pulmonary vascular resistance, the pressure in the pulmonary artery falls, causing a reversed blood flow through the ductus arteriosus.
10. **Closure of the ductus arteriosus** appears to be **initiated by the high O_2** tension of the arterial blood passing through it.
11. The presence of **vasodilator prostaglandins** is thought to be the reason for failure of the ductus arteriosus to close. The administration of **indomethacin,** which blocks prostaglandin synthesis, often leads to closure of the ductus in infants in whom it fails to close.

CUSHING PHENOMENON

- *Generally, cerebral blood flow to the brain is constant. Cerebral metabolic products (diminished O_2, elevated CO_2 and H^+) contribute to the control of cerebral blood flow locally in accordance with local metabolism.*
- *Cerebral circulation is maintained in hypertension by a combination of sympathetic vasoconstriction, hormonal vasoconstriction, and homeostatic autoregulation.*
- *However, the neurosurgeon Harvey Cushing noted that most of his patients with brain tumors who had cerebral ischemia also had increased systemic blood pressure with a simultaneous decrease in heart rate.*
- *This response, called **Cushing's phenomenon,** is caused by cerebral ischemic stimulation of vasomotor regions in the medulla that help maintain cerebral blood flow in the face of increased resistance caused by expanding intracranial tumors.*

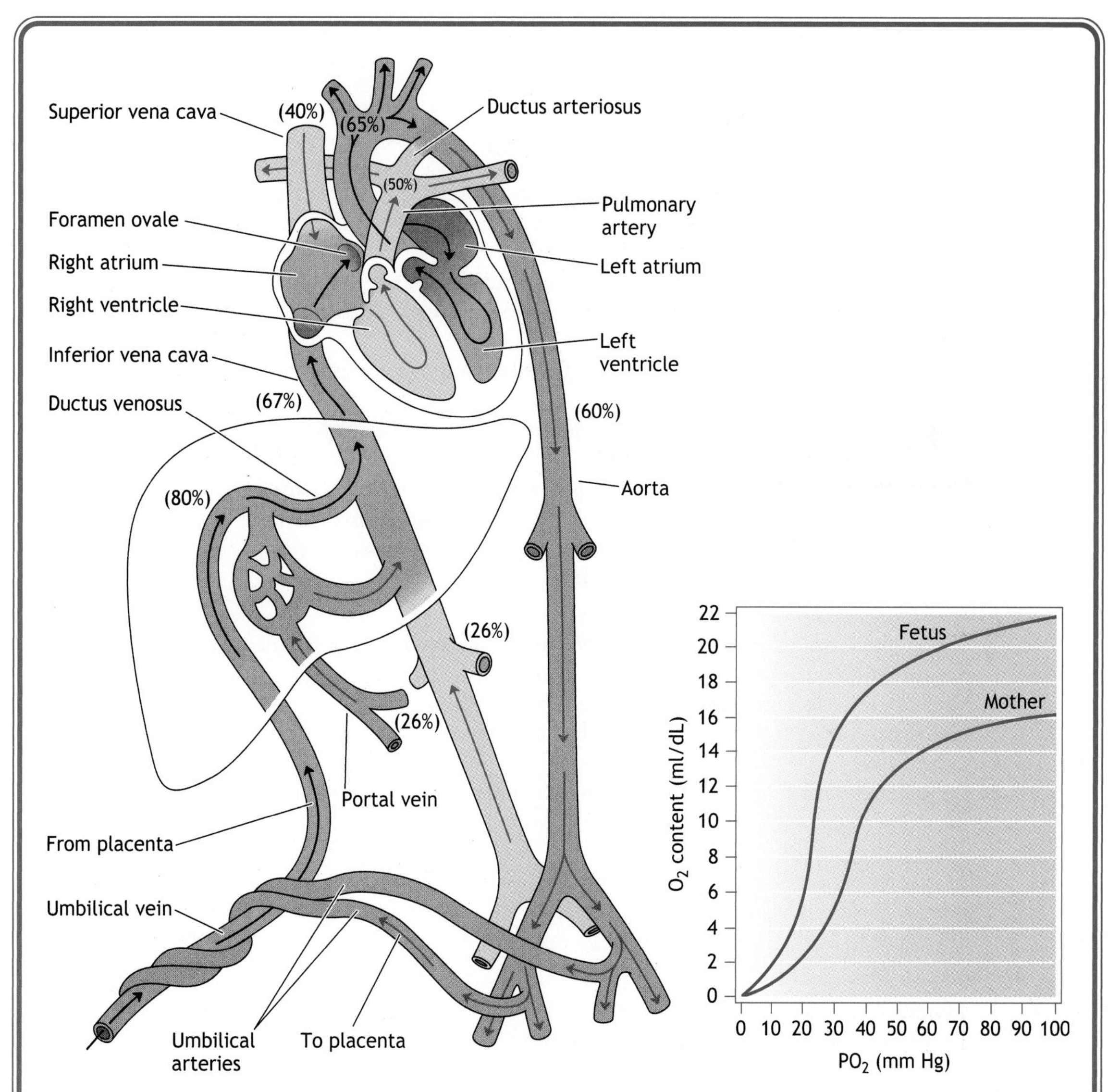

Figure 2–11. Fetal circulation at birth. The inset illustrates differences between fetal and adult hemoglobin. Fetal blood contains approximately 50% more hemoglobin than does maternal blood, and the oxyhemoglobin dissociation curve is shifted to the left for fetal hemoglobin. Thus, at low PO_2 blood levels, fetal hemoglobin can carry 20–50% more oxygen than maternal hemoglobin.

VIII. Integrative Function

A. Exercise and Decreased End-Diastolic Pressure

1. During exercise, **sympathetic outflow to the heart and blood vessels is increased.**
2. As **cardiac output** and **blood flow** to active muscles **increase** with progressive intensity of exercise, **splanchnic and renal blood flow decreases.**
3. **Blood flow** to the **myocardium increases,** whereas flow to the brain is unchanged.
4. The local accumulation of **vasodilator metabolites** relaxes the terminal arterioles, **and blood flow to active muscles may increase 20 times** above the resting level.
5. **O_2 consumption may increase** as much as **60 times,** whereas muscle blood flow increases up to 15 times.
6. **Increased venous return** is aided by the working skeletal muscles and by the muscles of respiration.
7. The large volume of blood returning to the heart is pumped through the lungs and out into the aorta so rapidly that **central venous pressure** (ie, **preload**) **remains** essentially **constant. In maximal exercise,** however, **right atrial pressure** and **end-diastolic volume increase.**

B. Muscle Contraction and Venous Valves

1. When one is **standing at rest,** the **venous valves are open** by virtue of the pressure difference between peripheral veins and the right atrium.
2. **Skeletal muscle contraction compresses the veins** so that increased pressure drives blood toward the thorax through the upper valves and closes the lower valves.
3. Immediately **after muscle relaxation,** the **pressure** in the previously contracted venous segment **falls,** and the reversed pressure gradient causes the upper valves to close.
4. This valve action is assisted by **inspiration,** which **raises abdominal venous pressure** while lowering thoracic venous pressure and increasing the pressure difference to facilitate venous return to the right heart.

C. Venous Return and Inspiration

1. The normal periodic activity of the respiratory muscles causes rhythmic variations in vena caval flow. Thus, **inspiration** constitutes an auxiliary pump to **promote venous return.**
2. The reduction in intrathoracic pressure during inspiration is transmitted to the lumina of the intrathoracic blood vessels. This reduction is accompanied by contraction of abdominal muscles and an increase in intra-abdominal venous pressures.
3. The reduction in central venous pressure during inspiration increases the pressure gradient between extrathoracic and intrathoracic veins, leading to **acceleration in venous return to the right atrium.**
4. **Sustained expiratory efforts** increase intrathoracic pressure and thereby **impede venous return.**
5. **Forced expiration** against a closed glottis (the **Valsalva maneuver**) regularly occurs during coughing, defecation, and heavy lifting.
   a. Intrathoracic pressures in excess of 400 mm Hg have been recorded during paroxysms of coughing.

b. Such pressure increases are transmitted directly to the lumina of the intrathoracic vessels, collapsing them.
c. After the Valsalva and opening the airway, intrathoracic pressures fall stimulating aortic baroreceptors and triggering a reflex vagal slowing of the heart.

D. Starling's Law and Position Change (Gravity) (Figure 2–12)

1. **Gravity causes a hydrostatic pressure difference when a difference in height occurs.**
2. If one **stands up suddenly,** blood pools in the legs, **venous return to the right ventricle drops,** and **stroke volume will** immediately **drop.**

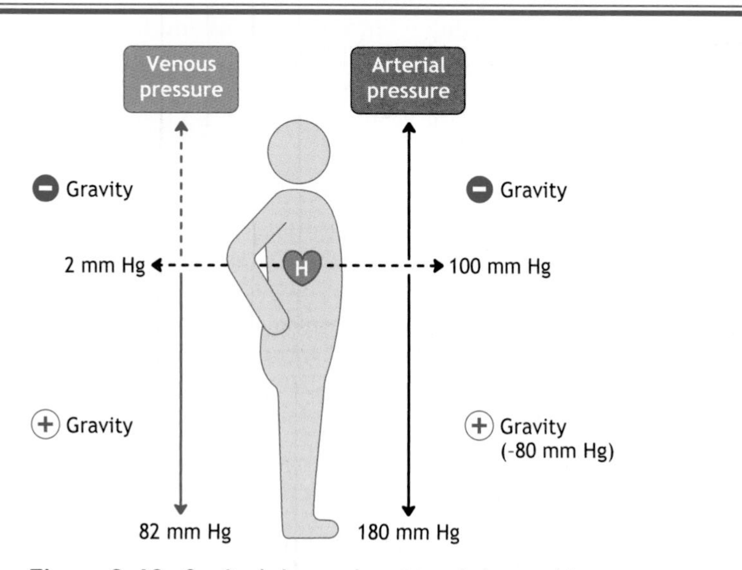

Figure 2–12. Starling's law and positional change. H stands for the heart, and everything in the figure is based on being above or below heart level. Below heart level, systemic arterial and venous pressures increase (+) equally, assuming there is no muscular activity. The pressure difference between veins and arteries is the same at the ankle and heart level. Gravity assists in this increase below heart level and pushes against the column of blood above the heart level. Elevated venous pressures indicate significant pooling of blood and decreased venous return. Because venous pressure is low at heart level, above-heart-level venous pressure becomes subatmospheric or negative (–). Arterial pressures also decrease progressively above heart level. When a person changes position from supine to upright, blood volume increases and dependent veins and venous pressure increase; circulating blood volume decreases and blood pressure decreases, particularly above heart level.

3. **Arterial baroreceptor reflex** compensation will minimize the fall in arterial pressure by **increasing myocardial contractility.**
4. Similar mechanisms occur with a person strapped to a **tilt board.** A **sudden head-down tilt** of the board **increases return of venous blood to the heart** from the legs, and **stroke volume** of the right ventricle **will rise.**

E. **Starling's Capillary Forces**

1. According to Starling's law, the rate and **direction of fluid movement is determined** by the balance of **hydrostatic and oncotic pressures.**
2. Pressures **favoring fluid filtration** include the **capillary hydrostatic pressure** and the **interstitial fluid oncotic pressure.**
3. Pressures **favoring fluid reabsorption** include the **capillary oncotic pressure** and the **interstitial fluid hydrostatic pressure.**
4. At the **arterial end of the idealized capillary,** the balance of pressures **favors filtration** (ie, a positive algebraic sum), whereas the **venous end favors reabsorption** (ie, a negative algebraic sum). The difference is caused by a fall in capillary blood pressure (P_c) from approximately 35 mm Hg at the arteriolar end to approximately 15 mm Hg at the venular end.

CAUSES OF EDEMA

Hydrostatic Forces

- *In **left ventricular failure** or **stenosis of the mitral valve,** pulmonary capillary hydrostatic pressure may exceed plasma oncotic pressure and cause **pulmonary edema.***
- ***With prolonged standing,** particularly associated with some elevation of venous pressure (eg, in pregnancy or with congestive heart failure), **filtration is greatly enhanced** and exceeds the ability of the lymphatic system to remove the capillary filtrate.*

Colloid Osmotic Forces

- *Plasma protein concentrations may decrease (eg, nephrotic syndrome) or increase (eg, from water deprivation, prolonged sweating, severe vomiting, or diarrhea) and thus alter the osmotic force and movement of fluid across the capillary.*

Lymphatic Drainage

- *Blockage of lymphatic vessels (eg, associated with pregnancy, or filariasis, a worm infestation) causes accumulation of interstitial fluid in the subcutaneous space.*

Properties of the Capillary Wall

- *Inflammatory conditions that cause the release of vasodilators that increase the number of open capillaries result in enhanced fluid filtration from capillary lumen to interstitium causing tissue swelling (eg, severe burns, toxins).*

CLINICAL PROBLEMS

A patient experiences a hemorrhage that lasts 30 minutes. At the end of that time, the mean arterial pressure has dropped from 90 to 75 mm Hg. The heart rate has increased from 70 to 150 beats/min, and the skin becomes cold.

1. At this time, it can be concluded that
   A. Capillary hydrostatic pressure is increased
   B. Interstitial fluid volume is increased
   C. Capillary colloidal osmotic pressure is increased
   D. Interstitial fluid pressure is increased
   E. The hematocrit has been decreased

2. Which of the following causes of brain hypoxia would most strongly stimulate the aortic and carotid chemoreceptors?
   A. Carbon monoxide poisoning
   B. Severe anemia
   C. Formation of methemoglobin
   D. A marked decrease in the pulmonary diffusing capacity
   E. Acute respiratory alkalosis

A 63-year-old man suddenly felt a crushing pain beneath his sternum. He became weak, was sweating profusely, and noticed his heart was beating rapidly. He called his physician, who made the diagnosis of myocardial infarction. The tests made at the hospital confirmed the doctor's suspicion that his patient had experienced a "heart attack." An ECG indicated that the SA node was the source of the rapid rate. Two hours after admission to the hospital, the patient suddenly became much weaker. His arterial pulse rate was only about 40 beats/min. An ECG revealed that the atrial rate was about 90 beats/min and that conduction through the AV conduction system was completely blocked, undoubtedly because the infarct affected the conduction system. Electrodes of an artificial pacemaker were inserted into the patient's right ventricle, and the ventricular rate was placed at a frequency of 75 beats/min. The patient felt stronger and more comfortable almost immediately. Soon after the occlusion, the interstitial fluid K^+ concentration rose substantially in the flow-deprived region.

3. This elevated extracellular K^+ concentration
   A. Increased the propagation velocity of the myocardial action potentials
   B. Decreased the repolarization refractoriness of the myocardial cells
   C. Increased the resting (phase 4) transmembrane potential to a less negative value
   D. Diminished the automaticity of the myocardial cells
   E. Decreased the likelihood of reentry dysrhythmias

A 70-year-old man complained of severe pain in his right leg whenever he walked briskly; the pain disappeared soon after he stopped walking. Angiography showed partial obstruction by large arteriosclerotic plaques about 3 cm distal to the origin of the right femoral artery. The mean pressure in the artery proximal to the obstruction was 100 mm Hg, and just distal to the obstruction it was 80 mm Hg. The blood flow in this artery was 300 mm Hg/mL/min. The mean venous pressure was 10 mm Hg.

4. The resistance to blood flow in the vascular bed perfused by the right femoral artery was
   A. 0.03 mm Hg/mL/min
   B. 0.30 mm Hg/mL/min
   C. 3.00 mm Hg/mL/min
   D. 3.33 mm Hg/mL/min
   E. 33.3 mm Hg/mL/min

A 33-year-old man complained about chest pain on exertion. He was referred to a cardiologist, who carried out a number of studies, including right- and left-sided catheterization. Among the data obtained during these studies were the findings that at the time of his initial examination the patient's mean aortic pressure was 93 mm Hg and his mean pulmonary artery pressure was 20 mm Hg.

5. These findings can be explained as follows:
   A. The patient's systemic vascular resistance was much greater than his pulmonary vascular resistance.
   B. The patient's aortic compliance was much greater than his pulmonary artery compliance.
   C. The patient's left ventricular stroke volume was much greater than his right ventricular stroke volume.
   D. The total cross-sectional area of the patient's pulmonary artery was much greater than the total cross-sectional area of the aorta.
   E. The duration of the rapid ejection phase of the patient's left ventricle exceeded the duration of the rapid ejection phase of the right ventricle.

A 44-year-old woman with severe cardiac failure caused by coronary artery disease was treated by cardiac transplantation. She recovered very well, and 1 month after surgery her cardiovascular function was entirely normal, even though her new heart was totally denervated. About 3 months after surgery, she developed a bleeding duodenal ulcer and was estimated to have lost about 600 mL of blood in 1 hour. Her physician treated the ulcer with dietary changes and antibiotics, and the ulcer was cured in about 2 weeks.

6. The acute blood loss from the patient's duodenal ulcer would be expected to
   A. Decrease central venous pressure and increase cardiac output
   B. Increase central venous pressure and decrease mean arterial pressure
   C. Decrease central venous pressure and decrease cardiac output
   D. Increase mean arterial pressure and decrease cardiac output
   E. Decrease central venous pressure and increase aortic pulse pressure

A 6-year-old boy was referred by his family physician to a pediatric cardiologist because of chronic fatigue, effort intolerance, and a heart murmur. On physical examination, the boy appeared slightly small for his age, had normal skin color, no clubbing of the fingers, and a harsh murmur throughout systole that was heard best in the fourth intercostal space to the left of the sternum but extended over the entire precordium. X-ray revealed an enlarged heart, especially the right ventricle. An ear oximeter showed normal oxygenation of arterial blood. Cardiac catheterization data were as follows:

Mean right atrial pressure	5 mm Hg
Right ventricular systolic pressure	30 mm Hg
Right ventricular diastolic pressure	3 mm Hg
Right atrial blood PO_2	40 mm Hg
Right ventricular blood PO_2	60 mm Hg

7. The patient was admitted to the cardiac surgery unit for repair of
   A. Coarctation of the aorta
   B. Interventricular septal defect
   C. Pulmonic stenosis
   D. Tetralogy of Fallot
   E. Patent ductus arteriosus

ANSWERS

1. The answer is E. Absorption of fluid, without protein or cells, decreases the hematocrit. Choice A is incorrect because with significant whole blood loss, the capillary hydrostatic pressure will be below normal, not increased. Choice B is incorrect because Starling's capillary forces favor reabsorption of interstitial fluid so that interstitial fluid volume would be decreased, not increased. Choice C is incorrect because with increased interstitial fluid reabsorption (without protein), capillary osmotic pressure is decreased, not increased. Choice D is incorrect because interstitial fluid volume is decreased, not increased.

2. D is correct. A significant decrease in pulmonary diffusing capacity will increase PCO_2 levels, which is the most important local vasodilator for cerebral vasodilation. In addition, the arterial PO_2 will be reduced (hypoxemia), which would provide an additional, albeit a lesser, stimulus to the chemoreceptors. Carbon monoxide (choice A) occupies O_2-binding sites on hemoglobin, thereby decreasing the O_2-binding capacity of hemoglobin and resulting in hypoxemia. PO_2 (dissolved oxygen only) must fall below 60 mm Hg, however, for a significant stimulation of peripheral chemoreceptors to occur. Severe anemia (choice B) is associated with reduced hemoglobin, but anemic persons almost never become cyanotic (> 5 g of deoxygenated hemoglobin) because there is not enough hemoglobin for 5 grams of it to be deoxygenated. Because the dissolved oxygen (PO_2) remains normal, the chemoreceptors would not be stimulated. Formation of methemoglobin (choice C) occurs when the ferrous iron in hemoglobin is converted to ferric iron, which is no longer able to react with oxygen, thereby decreasing the bound oxygen. Dissolved oxygen, however, remains normal; thus, peripheral chemoreceptors would not be stimulated. Acute respiratory alkalosis (choice E) is associated with decreased levels of CO_2 due to hyperventilation. Arterial PO_2 levels are normal, however,

with hyperventilation; therefore, the stimulus of the peripheral chemoreceptors would be less than with a decreased diffusing capacity.

3. The answer is C. With an accumulation of extracellular potassium, the transmembrane potential will become less negative. Interstitial potassium is elevated because of Na^+/K^+-ATPase pump hypoxic failure. Choice A is incorrect because elevation of extracellular potassium (via the Nernst equation) will partially depolarize the resting membrane potential, thereby decreasing phase 0 amplitude and decreasing the conduction velocity. Choice B is incorrect because repolarization of myocardial cells requires potassium extrusion from the interior of the cell. This is more difficult with elevated extracellular potassium. Choice D is incorrect because automaticity (firing frequency) is related to the time to reach threshold potential, which is decreased in the less negative cells closer to their threshold potential. Choice E is incorrect because a decrease in ventricular muscle conduction velocity will increase the likelihood of reentry dysrhythmias.

4. The answer is B. The resistance across the claudication is a portion of the total resistance in the limb. Hence, by the hemodynamic equivalent of Ohm's Law, $\delta P = Q \times R$, and $R = \delta P/Q$. $\delta P/Q = (100 - 10)/300 = 90/300 = 0.3$ mm Hg/mL/min.

5. The answer is A. $\delta P = Q \times R$. The right and left sides of the heart have the same cardiac output (Q). Systemic resistance exceeds pulmonary resistance (R), as indicated by the aortic pressure (93 mm Hg) versus the pulmonary artery pressure (20 mm Hg). Choice B is incorrect because compliance would influence pulse pressure, not average pressure. Choice C is incorrect because stroke volumes are nearly equivalent beat-to-beat, giving rise to equivalent ventricular outputs per minute. Choice D is incorrect because, with equivalent stroke volumes, cross-sectional area would determine velocity (Poiseuille), not pressure. Thus, cross-sectional area would not account for the pressure differences. Choice E is incorrect because the duration of ejection from the left ventricle is less, not more, than the duration of ejection from the right ventricle.

6. The answer is C. The patient's acute blood loss decreased the driving force for venous return, decreased central venous pressure, and decreased venous return, which means cardiac output decreased. Choice A is incorrect because acute blood loss would decrease peripheral venous pressure, thereby decreasing cardiac output, not increasing it. Choice B is incorrect because acute blood loss would decrease peripheral venous pressure, decrease driving pressure (ΔP) for venous return, and decrease venous return. Thus, central venous pressure would be decreased, not increased. Choice D is incorrect because reflex compensations would erase any influence of a reduction in cardiac output on arterial pressure. Thus, there would be no increase in pulse pressure. Choice E is incorrect because acute blood loss would decrease central venous pressure, decrease preload, decrease stroke volume, and decrease aortic systolic and pulse pressures, not increase aortic pulse pressure.

7. The answer is B. The murmur, the high right ventricular systolic pressure with a normal right atrial pressure, the elevated PO_2 of the right ventricular blood, and the absence of cyanosis indicate a left-to-right shunt through an interventricular septal defect. Coarctation (stenosis) of the aorta (choice A) would not be indicative of right ventricular enlargement. Only slight elevation of pulmonary systolic pressure rules out pulmonic stenosis (choice C), tetralogy of Fallot (choice D), and patent ductus arteriosus (choice E).

CHAPTER 3
RESPIRATORY PHYSIOLOGY

I. Lung Volumes and Capacities (Figure 3–1)

A. **Volumes are measured by spirometry except for residual volume and any volumes containing residual volume.**

1. **Tidal volume** (V_T) is the volume of air that moves into and out of the lung in each breath. Tidal volume is usually **about 500 mL.**
2. **Inspiratory reserve volume** (**IRV**) is the volume of air that can be inspired with maximum inspiratory effort, starting at the end of a normal inspiration. IRV is **2–3 L.**
3. **Expiratory reserve volume** (**ERV**) is the volume of air that can be expired with maximum expiratory effort, starting at the end of a normal expiration. ERV is **about 1.5 L.**
4. **Residual volume** (**RV**) is the volume of air remaining in the lungs (alveolar and dead space) after a maximum expiration. RV is about 1.5 L. RV cannot be measured from a simple spirometer record, because the **spirometer measures only changes in lung volume** and not the absolute amount of air in the lung.
5. **Total lung capacity** (**TLC**) is the sum of all four volumes (TLC = RV + ERV + V_T + IRV).
6. **Functional reserve capacity** (**FRC**) is the volume of air in the lungs at the end of a normal passive expiration (FRC = ERV + RV).
7. **Inspiratory capacity** (**IC**) is the maximum volume of air that can be inspired from the FRC (IC = V_T + IRV).
8. **Vital capacity** (**VC**) is the maximum volume of air that can be expired after a maximal inspiratory effort and is the sum of the TV, IRV, and ERV (VC = ERV + V_T + IRV).

B. **Ventilation** is the process that involves movement of air through the airways and into the alveoli.

1. **Total ventilation,** also called **minute ventilation, $\dot{V}_E$,** is defined as the volume of air entering and leaving the lungs per minute. Minute ventilation is equal to the tidal volume times the number of breaths per minute (average = 12/min).
2. Total ventilation, however, does not represent the inspired air that is available for gas exchange, because of the effect of the anatomic dead space.
3. The **anatomic dead space** (**V_D**)includes the conducting zone (airways that do not participate in gas exchange) that ends at the level of the terminal bronchioles. V_D averages 150 mL.

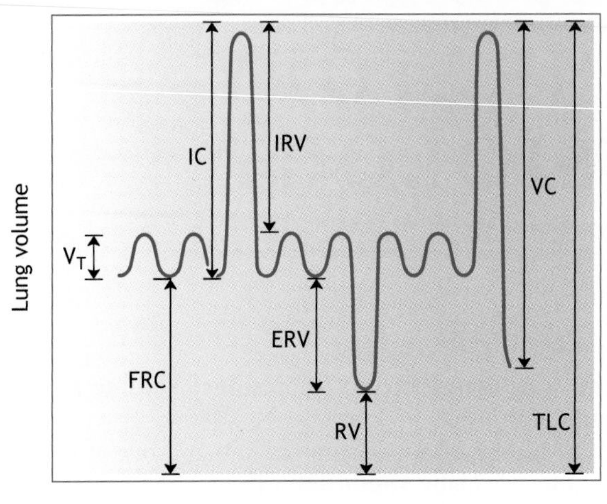

Figure 3–1. Subdivisions of lung volumes. The spirometry record shows an idealized summary of changes in lung volume during normal breathing, maximal inspiration, maximal expiration, and a maximal inspiration followed by a full expiration to the residual volume known as vital capacity (VC). The RV and FRC cannot be measured by spirometry. ERV, expiratory reserve volume; FRC, functional reserve capacity; IC, inspiratory capacity; IRV, inspiratory reserve volume; RV, residual volume; V_T, tidal volume; TLC, total lung capacity.

Respiratory Bronchioles = Transition zone

4. Alveolar ducts and alveolar sacs make up the **respiratory zone,** where significant gas exchange with blood occurs.
5. **At the end of expiration,** the anatomic dead space contains air that has come from the alveoli.
6. **During the initial phase of inspiration,** inspired air flows into the conducting zone, and anatomic dead space gas moves into the alveoli.
7. **At the end of inspiration,** the anatomic dead space is filled with humidified atmospheric air.
8. **With each tidal volume,** the volume of new air reaching the alveoli is $\mathbf{V_T - V_D}$. Similarly, the volume of alveolar air expelled with each breath is $\mathbf{V_T - V_D}$.
9. **Alveolar ventilation** ($\mathbf{\dot{V}_A}$) is defined as the total volume of alveolar air expired per minute. For example,

$$\dot{V}_A = (V_T - V_D) \times (\textit{breathing frequency})$$
$$= (500 \text{ mL} - 150 \text{ mL}) \times 12 = 4200 \text{ mL/min}$$

   a. **Increasing the depth of breathing** by 200 mL increases the total and alveolar ventilation by 200 mL.

   b. **Increasing the rate of breathing** will cause a greater increase in total ventilation than in alveolar ventilation.

10. **Forced vital capacity** (**FVC**) is the volume of air that can be forcibly expired after a maximal inspiration.
11. **Forced expiratory volume** (**FEV_1**) is the volume of air (normally around 80%) that can be expired in 1 second after a maximal inspiration.
    a. **The FEV_1**/FVC ratio is a pulmonary function test used to diagnose **obstructive** (eg, asthma) **and restrictive** (eg, fibrosis) disorders.
    b. In fibrosis the FEV_1/FVC ratio is increased, whereas the FEV_1/FVC ratio is decreased in asthma.

ASTHMA

- *Asthma is the **most common chronic pulmonary disease** occurring in 5–10% of the American population and is caused by gas-exchange abnormalities which produce hypoxemia (reduction of PO_2).*
- *Now recognized to be primarily an inflammatory disorder with bronchospasm secondary.*
- ***Clinical symptoms** include dyspnea (difficult breathing), cough, chest tightness, and wheezing with prolonged expiration.*
- *Severe attacks result in respiratory failure requiring tracheal intubation and mechanical ventilation.*
- *Asthma attacks can be triggered by exercise, pets, smoke, dust, or strong smells.*
- *Airway inflammation produces increased airway resistance and airflow obstruction. It also produces airway hyperresponsiveness, which causes bronchoconstriction.*
- ***Treatment** depends on the severity of attacks and involves bronchodilators and anti-inflammatory agents.*

II. Muscles of Breathing

A. Inspiration

1. The **diaphragm** is the major muscle of inspiration.
   a. This dome-shaped muscle is located between the thorax and the abdomen.
   b. It is innervated by phrenic nerves.
   c. The diaphragm moves down during inspiration and up during expiration.
   d. Quiet breathing is accomplished almost entirely by the diaphragm.
2. **External intercostals** are important muscles for active inspiration, for example, during exercise, singing, playing wind instruments, and sighing.
   a. These muscles are located between the ribs and are oriented such that contraction elevates the ribs and increases thickness of the thoracic cage, thereby drawing air into the lungs.
   b. They are innervated by intercostal nerves that come from the spinal cord at the level of the rib attached to a given intercostal muscle.
3. The **accessory inspiratory muscles** are the scalene and sternomastoid muscles and the alae nasi (used in nostril flaring).
4. **Contracting the diaphragm and external intercostal muscles increases the volume of the thorax, producing an inspiration.**

B. Expiration

1. Relaxing the muscles of inspiration produces a quiet passive expiration.
2. The **abdominals** are the main muscles of expiration. Contraction of these muscles opposes the action of the diaphragm, that is, tending to push the diaphragm upward.

3. The **internal intercostals** oppose action on the external intercostals. They are oriented so that contraction tends to pull the rib cage down and decreases the anterior-posterior thickness of the thorax.

C. Forces Acting on the Lungs

1. **Lung recoil** refers to forces that develop in the lung wall during expansion.
   a. Recoil increases as the lung enlarges.
   b. Recoil always acts to collapse the lung.
2. **Intrapleural pressure** (**P_{IP}**) (also called pleural pressure, or **P_{IP}**) is the pressure in the thin film of fluid between the lung and chest wall (Figure 3–2).
   a. The balance between the outward elastic recoil of the chest wall and the inward elastic recoil of the lungs generates a subatmospheric intrapleural pressure (P_{IP}) (~ −5 cm H_20).
   b. Negative subatmospheric pressures act to expand the lung, whereas positive pressures act to collapse the lung.
   c. It is best to think of P_{IP} as the intrathoracic pressure which is the pressure throughout the lung except inside blood vessels, lymphatics, or airways.
   d. When P_{IP} exceeds recoil forces the lungs expand.
   e. When recoil forces exceed P_{IP} the lungs decrease in volume.
3. **Alveolar pressure** (**P_A**) is the pressure of the alveolar air (see Figure 3–2).
   a. P_A drives airflow into and out of the lungs.
   b. If P_A equals 0 (ie, no airflow with the glottis open), then P_A is the same as atmospheric pressure.
   c. P_A is less than 0 during inspiration; P_A is greater than 0 during expiration.
4. **Transpulmonary pressure** (**P_{TP}**) is the difference between the pressure inside the lung (alveolar pressure P_A) and the pressure outside the lung (intrapleural pressure P_{IP}). P_{TP} determines the degree of inflation of the lung.
5. **Pneumothorax** is the presence of atmospheric air in the pleural space.
   a. With no vacuum to counter their elastic recoil, the lungs collapse.

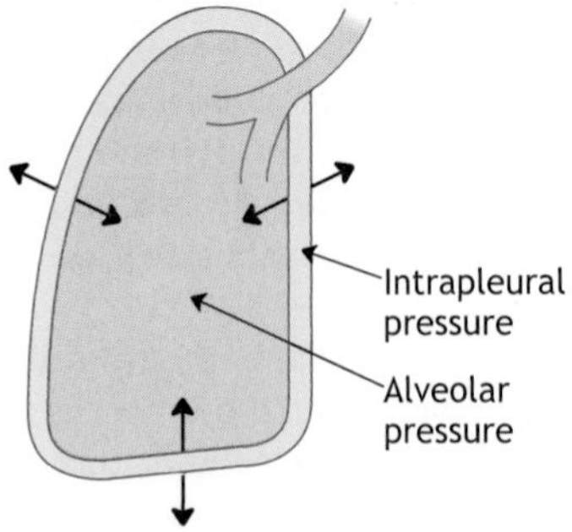

Intrapleural pressure

Alveolar pressure

Figure 3–2. Alveolar and intrapleural pressures during normal breathing. Intrapleural pressure remains negative during inspiration and expiration. Alveolar pressure is negative during inspiration and positive during expiration.

b. This condition is known as **atelectasis.**
c. The chest wall expands.

Atelectasis = collapse of the lung!

III. Lung Compliance

A. **Compliance** (C_L) is the stretching of the lungs and is calculated as follows (Figure 3–3):

$$C_L = \frac{\Delta V}{P_{TP}},$$

where
ΔV = change in lung volume
P_{TP} = transpulmonary pressure

B. Compliance is the change in lung volume per unit change in airway pressure. For example,

$$C_L = \frac{\Delta V}{\Delta_{P_{TP}}} = \frac{1000 \text{ mL}}{5 \text{ cm } H_2O} = 200 \text{ mL/ cm } H_2O$$

→ Normal value

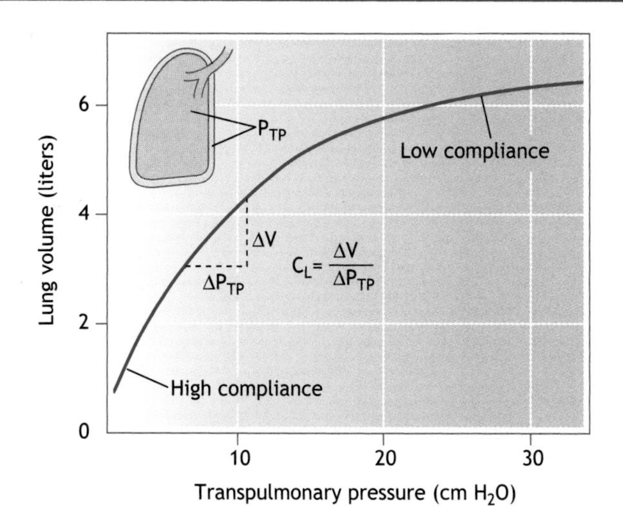

+10 ↑expansion ↓C

+2

Figure 3–3. Transpulmonary pressure (P_{TP}) is determined by subtracting intrapleural pressure (P_{PL}) from alveolar pressure (P_A). Thus, P_{TP} is greater in the upper regions of the lung, where P_{PL} is more negative and holds the lungs in a more expanded position. The upper regions of the lungs also have greater volumes than the lower regions. Further increases in volume per unit increase in P_{TP} are smaller in the upper than lower regions of the lungs because the upper expanded lung is stiffer (ie, less compliant).

C. High C_L means more air will flow for a given change in pressure.

D. Low C_L means less air will flow for a given change in pressure.

E. If P_{TP} becomes more negative, more air will flow into the system, and if P_{TP} becomes more positive more air will flow out of the system.

F. C_L is an indicator of the effort required to expand the lungs to overcome recoil.

G. **Compliant lungs** have low recoil, whereas stiff lungs have a large recoil force (Figure 3–4).

H. The **pressure-volume curve** is not the same for inspiration and expiration; this difference between inflation and deflation paths is called **hysteresis,** and is because a greater pressure difference is required to open a previously closed airway than to keep an open airway from closing.

IV. Components of Lung Recoil

A. The **collagen and elastic fibers** of the lung tissue provide elastance, which is the reciprocal of compliance.

B. **Surface tension forces** at the liquid-air interface of the airway accounts for at least half of the elastic recoil of the lungs.

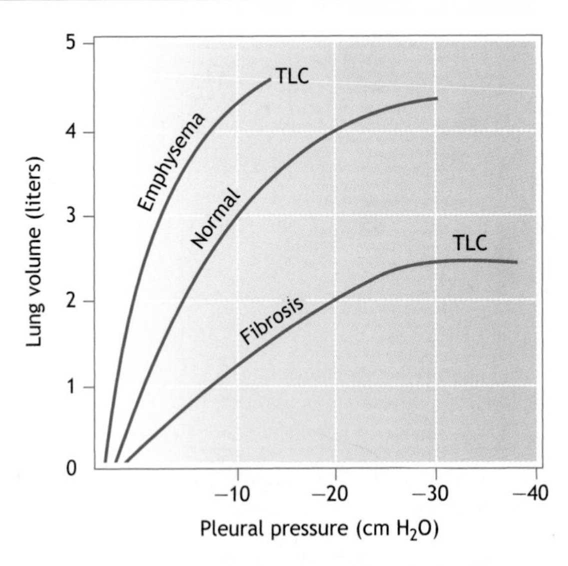

Figure 3–4. Static compliance curves are shown for normal and pathologic states. In fibrosis (lower curve) the lungs are stiff and less compliant and have increased alveolar elastic recoil force. Emphysema (upper curve) increases the compliance of the lungs and decreases alveolar elastic recoil forces because the alveolar septal tissue that opposes lung expansion is destroyed. Abbreviation key: TLC, total lung capacity.

1. The fluid lining the alveoli contains **surfactant,** a surface-tension-lowering agent, composed of a mixture of lipids and proteins.
   a. Half of the lipid is dipalmitoylphosphatidylcholine (DPPC) and albumin and secretory IgA make up about half of the protein.
2. Surfactant has three main functions.
   a. It **lowers surface tension forces** in the alveoli, which reduces lung recoil and **increases compliance.**
   b. The reduction in surface tension forces in small alveoli **decreases their tendency to collapse.**
   c. It also **reduces capillary filtration forces,** which decreases the risk of pulmonary edema.

RESPIRATORY DISTRESS SYNDROME (RDS)

- ***Neonatal RDS*** *is due to a deficiency of surfactant, seen in infants born prematurely.*
- *Surfactant synthesis and secretion are low until immediately before birth.*
- *Lung washings from infants with RDS exhibit a* ***high surface tension.***
- ***Prematurity and maternal diabetes*** *are risk factors.*
- *The tendency for small alveoli to collapse (called* ***atelectasis****) is increased, and reinflating the lungs is difficult after collapse.*
- *The increased negative intrathoracic pressure needed to maintain lung volume promotes capillary filtration and pulmonary edema.*

V. Airway Resistance

A. The **rate of airflow** is inversely proportional to **airway resistance:**

$$\dot{V} = \frac{\Delta P}{R} = \frac{P_A - P_B}{R}$$

where
$\dot{V}$ = flow rate (L/s)
ΔP = driving pressure
P_A = alveolar pressure (mm Hg)
P_B = barometric pressure (mm Hg)
R = airway resistance (R units)

The more negative the intrapleural pressure (eg, during inspiration), the lower the airway resistance.

B. **According to Poiseuille's equation,**

$$resistance \propto \frac{1}{r^4},$$

where
r = radius of the airway

Thus, a strong relationship exists between resistance and the radius of the airway.

C. The following factors influence airway resistance:
1. **Stimulation of parasympathetic nerves** and histamine produces **bronchoconstriction** and increased airway resistance.

2. **Stimulation of sympathetic nerves** or circulating catecholamine produces **bronchodilation.**
3. **Low lung volumes** are associated with **increased airway resistance,** whereas **high lung volumes** are associated with **decreased resistance.**
4. Breathing a **high-density gas increases resistance** to airflow, whereas breathing a low-density gas decreases resistance to airflow.
5. The first and second (ie, **medium-sized**) **bronchi** represent **most of the airway resistance.**
   a. The smallest airways contribute only slightly to total airway resistance.

RESTRICTIVE LUNG DISEASE

- *An example of restrictive lung disease is* ***fibrosis.***
- *Patients exhibit* ***reduced lung compliance*** *(poor expansion or inspiration) and* ***increased elastic recoil*** *(increased recoil or expiration).*
- *Restrictive lung disease can be caused by inhalation of asbestos fibers (****asbestosis****) or silica particles (****silicosis****).*
- *Total lung capacity is smaller, but the volume is expired more quickly and completely than normal.*

OBSTRUCTIVE LUNG DISEASE

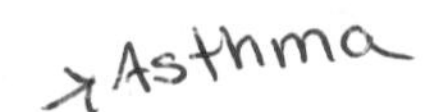

- *Examples of* ***chronic obstructive pulmonary disease*** *(****COPD****) include* ***emphysema*** *(****pink puffer****) and* ***chronic bronchitis*** *(****blue bloater****).*
  *–**Emphysema** is caused by permanent enlargement of distal airspaces leading to progressive dyspnea and irreversible obstruction.*
  –The patient with emphysema is likely to be thin and has diminished breath sounds and increased lung compliance but no cough.
  *–**Chronic bronchitis** results in an inflammatory reaction resulting in mucosal thickening and mucus hypersecretion leading to diffuse obstruction and cyanosis.*
  –The patient with chronic bronchitis is likely to be obese with some cyanosis (hence, the term blue bloater) and has productive cough, wheezing, severe respiratory acidosis, and normal lung compliance.
- *Chronic bronchitis involves central regions of the lung, whereas emphysema involves more distal regions.*
- *Both disorders are characterized by increased airway resistance due to destruction of lung parenchyma or irreversible damage to conducting pathways.*
- *Cigarette smoking is the most* ***common cause of COPD.***
- ***Total lung capacity is normal or larger than normal,*** *but a smaller than normal volume is slowly expired.*
- *Breathing is rapid and shallow with a prolonged expiration against pursed lips to keep the airways from collapsing.*
- *Table 3–1 compares obstructive and restrictive forms of lung disease.*

VI. Gas Exchange and Oxygen Transport

A. **Partial pressure** equals the total pressure times the fractional gas concentration.

B. **Assuming that total pressure is atmospheric** (760 mm Hg) and the fractional concentration of O_2 is 0.21, then

$$P_{O_2} = 0.21 \times 760 = 160 \text{ mm Hg}$$

Table 3–1. Obstructive and restrictive forms of lung disease.

Variable	Obstructive Form (eg, emphysema)	Restrictive Form (eg, fibrosis)
Total lung capacity	⇑	⇓
Forced expiratory volume (FEV_1)	⇓	⇓
Forced vital capacity (FVC)	⇓	⇓
FEV_1/FVC	⇓	⇑ or normal
Peak flow	⇓	⇓
Functional residual capacity	⇑	⇓
Residual volume	⇑	⇓

C. The **partial pressure** of humidified inspired air is calculated as follows:

$$PI\,\text{gas} = F\,\text{gas}\,(P_{atm} - PH_2O),$$

where
P_{atm} = atmospheric pressure
PI gas = partial pressure of inspired gas
PH_2O = partial pressure of H_2O vapor
F gas = concentration of gas

The partial pressure of H_2O at 37° is 47 mm Hg. Thus,

$$PI_{O_2} = 0.21\,(760 - 47) = 150\text{ mm Hg}$$

D. **Because 2% of cardiac output** bypasses the pulmonary circulation via a **physiologic shunt,** the PO_2 of arterial blood is lower than that of alveolar air.

E. **Physically dissolved oxygen (O_2)** consists of free O_2 molecules in solution. O_2 is also carried in blood bound to **hemoglobin (Hb)**.
   1. Hb consists of two α and two β subunits each of which has an iron containing "heme" and a polypeptide "globin".

F. **The amount of physically dissolved O_2** is directly proportional to the **PO_2** but is too small to meet the metabolic demands of the body. The units of concentration for a dissolved gas are mL gas per 100 mL blood.

G. **At body temperature,** blood equilibrated with a normal PO_2 (~ 100 mm Hg) contains only 0.3 mL O_2/100 mL blood (0.3 vol%), which is not enough to supply the needs of the tissues.

H. **Saturation is the percentage of Hb-binding sites occupied** by O_2.

1. **Each gram of Hb** has an oxygen capacity of 1.34 mL O_2, and because 100 mL of blood contains 15 g Hb, completely oxygenated blood contains approximately 20 mL O_2 (1.34 mL $O_2 \times$ 15 g Hb/100 mL).
2. **Thus, the oxygen capacity** of Hb in blood is approximately 20 mL O_2/100 mL of blood or 20 vol%.

I. Physiologic implications of the **oxyhemoglobin dissociation curve** include the following (Figure 3–5):

1. **Hb combines** rapidly and reversibly **with O_2** to form oxyhemoglobin.
2. The hemoglobin-O_2 dissociation curve has a sigmoid shape because oxygenation of the first heme group of the Hb molecule increases the affinity of O_2 for the other heme groups.
3. The **O_2 capacity** is the maximum amount of O_2 that can be bound to Hb and is determined by the Hb concentration in blood.
4. The **O_2 content** is the total amount of O_2 carried in the blood whether bound or dissolved in solution.
5. Figure 3–6 shows the **dissociation curve** as a function of partial pressure for two different amounts of Hb. The Hb concentration in normal blood is about 15 g/100 mL. The maximal amount of O_2/100 mL (98% saturation) in combination with Hb is 20.1 mL O_2/100 mL (1.34 mL $\times$ 15). The amount of dissolved O_2 is a linear function of the PO_2 (0.003 mL/100 mL blood/mm Hg PO_2).
   a. **In curve A,** the total amount of O_2 bound to Hb at 98% saturation is 19.7 mL O_2/100 mL blood. With the 0.3 mL/100 mL of dissolved O_2 added, the total O_2 content is approximately 20 mL O_2/100 mL of blood.

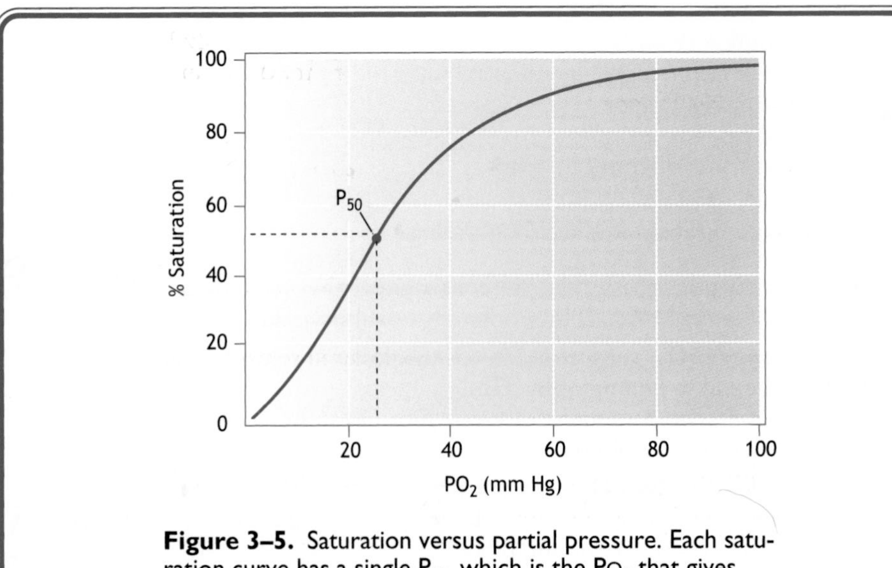

Figure 3–5. Saturation versus partial pressure. Each saturation curve has a single P_{50}, which is the PO_2 that gives 50% saturation. Normal P_{50} = 26 mm Hg.

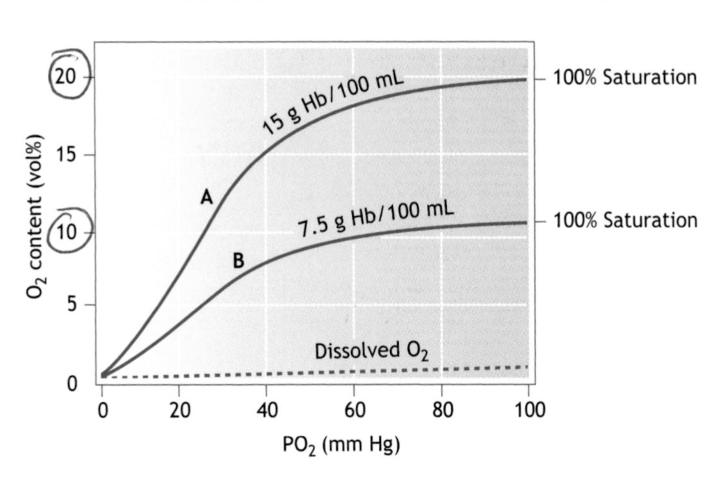

Figure 3–6. O_2 content versus partial pressure for two different hemoglobin (Hb) concentrations. Curve A represents normal Hb levels in blood (15 g/100 mL). Curve B represents a reduced concentration of Hb in blood (7.5 g/100 mL). The main effect of the lower Hb concentration is a reduced carrying capacity of the blood. Thus, in curve B, the total amount of O_2/100 mL of blood is around 10 mL O_2/100 mL, instead of the normal 20 mL O_2/100 mL.

   b. **In curve B,** the Hb is also 98% saturated, but this blood contains only 7.5 g Hb/100 mL blood. The total amount of O_2 bound to Hb is only 10 mL O_2/100 mL blood. Because of the lower amount of Hb, the amount of O_2 is about half that of normal blood.

J. Several factors influence the oxyhemoglobin dissociation curve (Figure 3–7).
   1. **Shifts to the right** occur when the affinity of Hb-binding sites for O_2 is decreased and it is easier for tissues to extract oxygen.
      a. Causes of this shift include increased CO_2 (Bohr effect), increased H^+ (decreased pH), increased temperature, and increased 2,3-diphosphoglycerate (2,3-DPG) all of which are characteristic of metabolically active tissues.
      b. 2,3-DPG reduces the O_2 affinity of adult but not fetal hemoglobin
      c. **Anemia** is characterized by a reduced Hb concentration in blood and decreased arterial oxygen content.
   2. **Shifts to the left** occur when there is increased affinity of Hb for O_2 and it is more difficult for tissues to extract oxygen.
      a. Causes of this shift include decreased temperature, decreased P_{CO_2}, decreased H^+ (increased pH), and decreased 2,3-DPG.
      b. **Stored blood loses 2,3-DPG and fetal Hb,** both **shift the curve to the left.**
      c. **Polycythemia** (increased number of red blood cells in blood) is characterized by a higher than normal concentration of Hb in the blood, a shift to

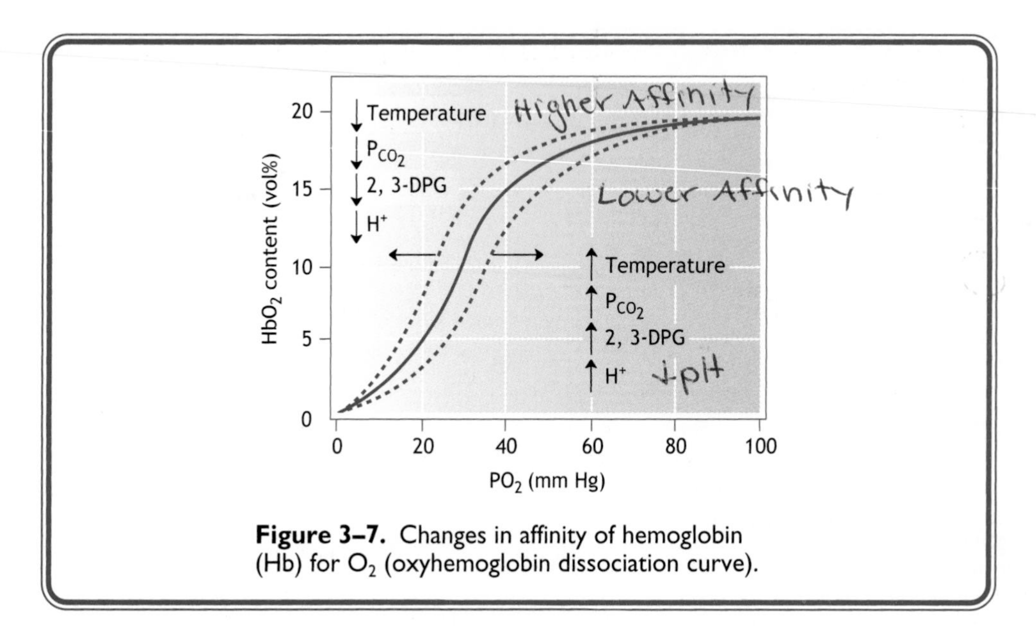

Figure 3–7. Changes in affinity of hemoglobin (Hb) for O_2 (oxyhemoglobin dissociation curve).

the left in the oxyhemoglobin dissociation curve, and increased arterial oxygen content.

CARBON MONOXIDE POISONING

- ***Carbon monoxide*** *(CO) has a much* ***greater affinity*** *(200–300 times higher)* ***for Hb*** *than does O_2 which prevents Hb from releasing O_2 to the tissues.*
- *Breathing high levels of CO for long periods causes the amount of CO dissolved in plasma to be essentially zero.*
- *The* ***oxyhemoglobin dissociation curve shifts to the left.*** ↑ affinity
- ***Arterial PO_2 is normal, but O_2 saturation of Hb is significantly decreased.***
- *Effects of maternal CO poisoning are magnified on the developing fetus.*
- *In the presence of 10% HbCO in both maternal and fetal blood, fetal arterial Po_2 is reduced because the maternal venous Po_2 is lower.*
- *Because O_2 must be released from maternal blood in the placenta before it can reach the fetus,* ***both the arterial PCO_2 and O_2 content of fetal blood are lowered by HbCO.***

VII. Carbon Dioxide Transport

A. $\mathbf{CO_2}$ is an important end-product of aerobic cellular metabolism and is, therefore, continuously produced by body tissues.

B. **After formation, $\mathbf{CO_2}$** diffuses into the venous plasma, where it is 24 times more soluble than O_2 and then passes immediately into red blood cells.

C. $\mathbf{CO_2}$ is carried in the plasma in three forms:

1. **Ninety percent is in the form of bicarbonate** from reaction with H_2O to form carbonic acid in the red blood cells, which dissociates into hydrogen and bicarbonate.
2. **Five percent is dissolved $\mathbf{CO_2}$**, which is free in solution.

3. **Five percent is in the form of carbaminohemoglobin,** which is CO_2 bound to hemoglobin.

D. CO_2 transport is dependent on carbonic anhydrase, the Cl-HCO_3 exchanger, and hemoglobin.

E. The uptake of Cl^- in exchange for HCO_3^- in red blood cells is called the **chloride shift** the HCO_3^- helps to maintain electrical neutrality and is transported to the lungs (Figure 3–8).

F. **Inside the red blood cell, deoxyhemoglobin** is a better buffer for H^+, and H^+ binding by deoxygenated Hb occurs in peripheral tissues where CO_2 is high.

G. **The enhancement of CO_2** binding to deoxygenated Hb at the venous end of capillaries leads to the formation of bicarbonate in red blood cells.

H. **In the lung, the reaction in the pulmonary capillaries** is in the opposite direction: The high PO_2 in the lung causes the blood to dump CO_2 into the alveolus for expiration, and HCO_3^- enters the red blood cells in exchange for Cl^- and combines with H^+ to form H_2CO_3.

I. In summary, **CO_2 entering the red blood cells causes a decreased pH that facilitates O_2 release.** In lungs, O_2 binding to Hb lowers the CO_2 capacity of blood by lowering the amount of H^+ bound to Hb.

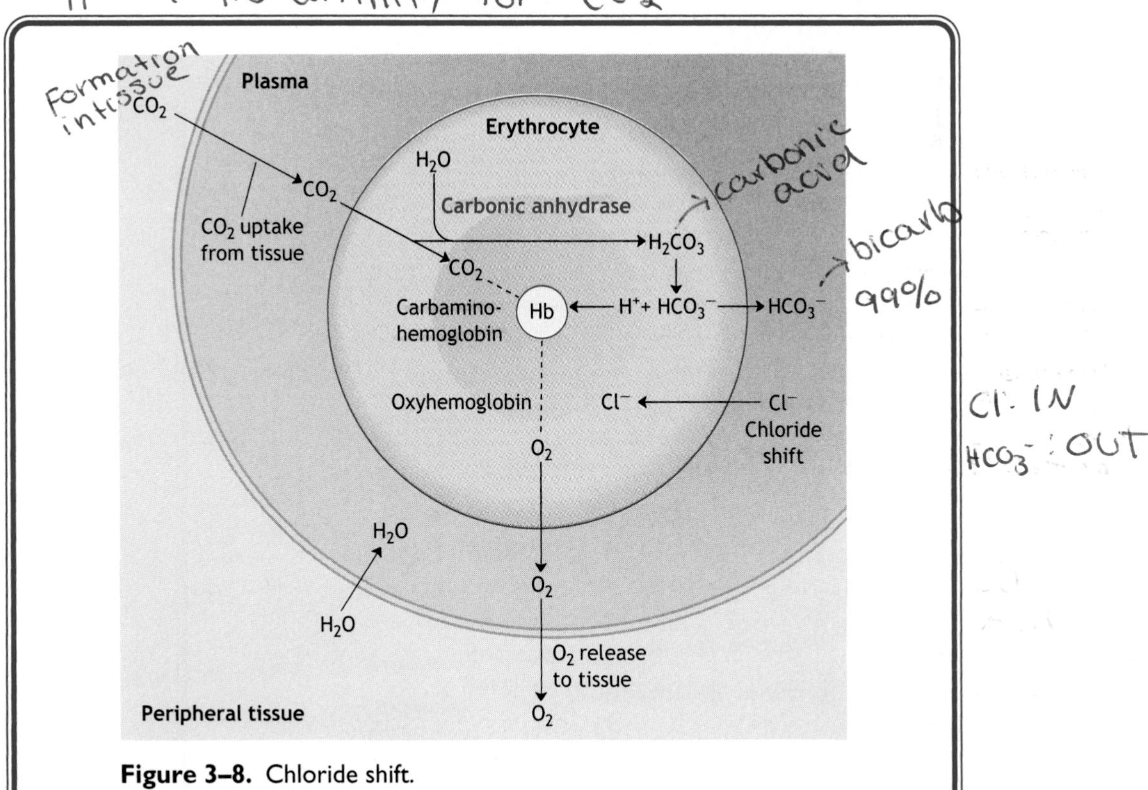

Figure 3–8. Chloride shift.

VIII. Respiration Control (Figure 3–9)

A. Brain stem centers drive, respiratory muscle activity rhythmically and subconsciously.

1. **Two main groups of respiratory neurons,** the dorsal respiratory group and the ventral respiratory group, comprise the **medullary respiratory center.**
2. The dorsal and ventral respiratory groups fire in phase with the respiratory cycle.
   a. The **dorsal respiratory group** processes sensory input and is responsible for the inspiratory respiratory rhythm; input comes from the vagus and glossopharyngeal nerves and output is via the phrenic nerve to the diaphragm.
   b. The **ventral respiratory group** is primarily motor and innervates both inspiratory and expiratory muscles but is primarily responsible for expiration. It becomes active only during exercise.

B. The **apneustic center** in the lower pons has an intrinsic rhythm and when stimulated promotes prolonged inspirations.

1. **Apneustic breathing** is an abnormal breathing pattern characterized by prolonged inspirations alternating with short periods of expiration.

C. The **pneumotaxic center** is located in the upper pons and has an inhibitory influence on the **apneustic center. If the connection** between the pneumotaxic center and apneustic center **is cut, apneustic breathing occurs.**

D. Apneustic and pneumotaxic centers modulate respiration but are not essential for normal respiratory output.

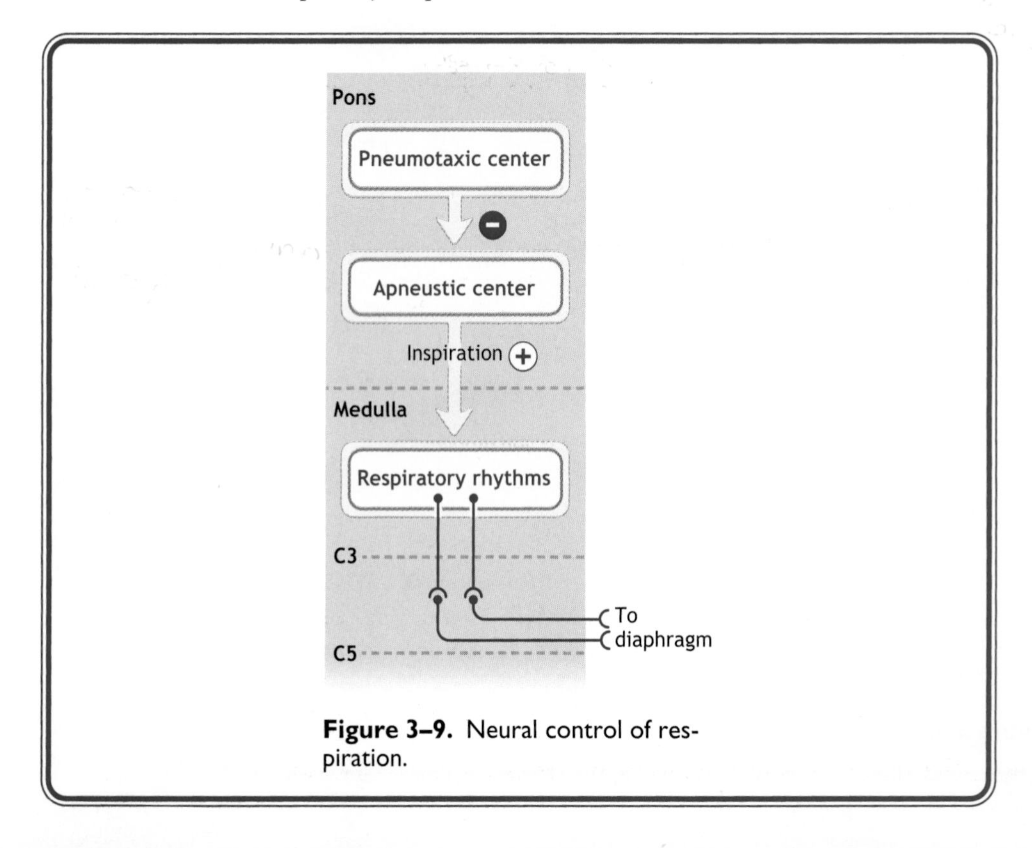

Figure 3–9. Neural control of respiration.

E. **Central chemoreceptors** are located in the ventrolateral medulla and are the **most important chemoreceptors in the regulation of normal breathing.**
  1. The blood-brain barrier separates the central chemoreceptors in the medulla from arterial blood.
  2. Because the blood-brain barrier is permeable to CO_2, increases in P_{CO_2} and $[H^+]$ stimulate breathing and decreases in P_{CO_2} and $[H^+]$ inhibit breathing.
  3. The **receptors are stimulated by cerebrospinal fluid** (**CSF**) **$[H^+]$ and CO_2** because they are sensitive to CSF pH.
  4. Therefore, the **primary drive for ventilation is CO_2** (H^+) on the central chemoreceptors.

F. **Peripheral chemoreceptors** are found in small bodies in two locations and respond to hypoxia, hypercapnia, and acidosis.
  1. **Carotid bodies** (the more important of the two types) are found at the bifurcation of the common carotid arteries (ie, near the carotid sinus). **Aortic bodies** are found near the aortic arch.
     a. The **glomus cell** is the chemosensor for hypoxia, hypercapnia, and acidosis in the carotid and aortic bodies.
     b. The common transduction pathway is the inhibition of K^+ channels resulting in a rise in glomer cell calcium concentration and the release of neurotransmitters (acetylcholine, norepinephrine, etc.) that control firing of nerve endings.

G. Higher brain centers coordinate ventilation with other behaviors and can override brain stem control of breathing.
  1. Coordination of voluntary behaviors that use respiratory muscles (eg, speaking).
  2. Changes in respiration via the limbic system (eg, fear, rage).
  3. Balancing conflicting gas-exchange demands and behavior (eg, playing a wind instrument).

CHEYNE-STOKES BREATHING

- *Cheyne-Stokes breathing refers to periods of* ***hyperventilation alternating with*** *periods of* ***apnea.***
- *Although the exact cause of this breathing pattern is not known, it occurs in patients with central nervous system lesions, cardiovascular disease, and normal people sleeping at high altitude.*
- *During periods of apnea, brain P_{CO_2} levels increase to stimulate ventilation maximally.*
- *Continued hyperventilation reduces alveolar P_{CO_2} below the desired set point, thereby inhibiting respiration.*
- *Thus, chemoreceptors receive information too late to regulate ventilation properly.*

IX. Pulmonary Blood Flow

A. Pressures within the Pulmonary Circuit

  1. The **most important difference** between the pulmonary and systemic circulations **is the low blood pressure** in the pulmonary arteries. The pulmonary arterial systolic pressure is approximately 22 mm Hg, whereas the left ventricular systolic pressure is around 120 mm Hg.
  2. **The pulmonary circulation** is a low-resistance circuit that must accommodate the entire cardiac output at rest and during exercise.
  3. When pulmonary arterial pressure increases, vascular resistance decreases for two reasons:

PT, RL

a. Increased pressure increases the caliber (**distention**) of the arteries.
b. Increased pressure causes more capillaries to open (**recruitment**).

B. Effects of Gravity on Blood Flow

1. Because of the low blood pressures in the pulmonary circulation, **gravity has a large effect on blood flow** to different parts of the lung (Figure 3–10).
   a. In an upright subject, the effect of gravity causes blood flow to be larger at the base than at the apex. Ventilation is also larger at the base than at the apex.
   b. Although the **base receives the greatest ventilation,** it does not match the very high blood flow. Thus, the base is an underventilated region, in which the $\frac{\dot{V}_A}{\dot{Q}}$ ratio is less than 0.8.
   c. Even though the **apex receives the lowest ventilation,** it is too high for the low blood flow. Therefore, the apex can be considered an overventilated region, in which the $\frac{\dot{V}_A}{\dot{Q}}$ ratio is greater than 0.8.

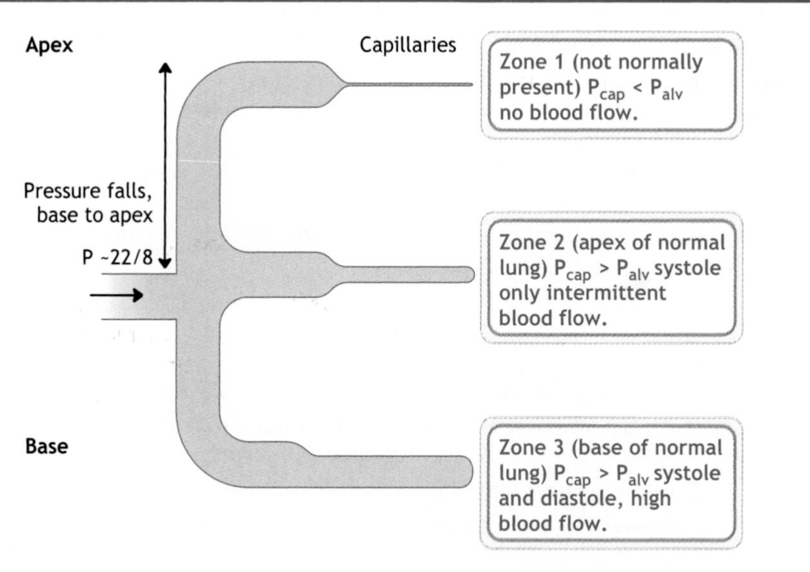

Figure 3–10. Circulation divisions in a vertical lung. The left side of the figure illustrates experimental data from upright animal lungs showing that pulmonary arterial pressure falls from the base of the lung to the apex. The amount of blood flow is illustrated by the line thickness. In zone I, no blood flow (thin line) occurs because the alveolar pressure is higher than the pulmonary arterial pressure. In normal persons with adequate cardiac output, there is no zone I because the pulmonary arterial pressure is greater than the alveolar pressure. In zone 2, the pulmonary arterial pressure is greater than the alveolar pressure; therefore, blood flow occurs (medium-width line), but the alveolar pressure is greater than the pulmonary venous pressure. In zone 3, the pulmonary arterial and venous pressures are both greater than the alveolar pressure; therefore, blood flow is maximal (thick line).

d. An overventilated lung unit acts like dead space, whereas an underventilated lung unit acts like a pulmonary shunt.

2. Regional blood flow in the lungs has been separated into three zones (see Figure 3–10).

C. Hypoxic Vasoconstriction

1. **A decrease in alveolar Po_2** produces a local vasoconstriction of pulmonary arterioles, thereby lowering blood flow to that part of the lung.
2. **In other systemic organs, hypoxia results in vasodilation** of arterioles.

D. Pulmonary Edema

1. For normal respiratory function, it is crucial that the alveoli do not accumulate fluid.
2. A small amount of fluid moves into peribronchial and perivascular spaces each day but is removed by lymphatic vessels.
3. If net fluid movement out of the pulmonary capillaries exceeds the ability of the lymphatic system to remove it, a net fluid accumulation, or edema, occurs.
4. Severe alveolar edema occurs when accumulated fluid in alveoli impairs normal gas exchange.
5. The two **causes of pulmonary edema** are
   a. Increased capillary permeability
   b. Increased pulmonary blood pressure due to hypoxic vasoconstriction, left heart failure, or loss of surfactant

E. Shunts

1. In **an absolute right-to-left shunt,** venous blood is delivered to the left side of the heart without contacting ventilated alveoli; this shunt produces **hypoxemia** (Figure 3–11).
   a. The shunt **results in a decrease in arterial Po_2 and widening of the Po_2 systemic alveolar-arterial (A-a) difference.**
   b. With a significant pulmonary shunt (such as occurs in regional atelectasis), breathing 100% O_2 does not result in a significant increase in systemic arterial Po_2, leading to a diagnosis of a pulmonary right-to-left shunt.

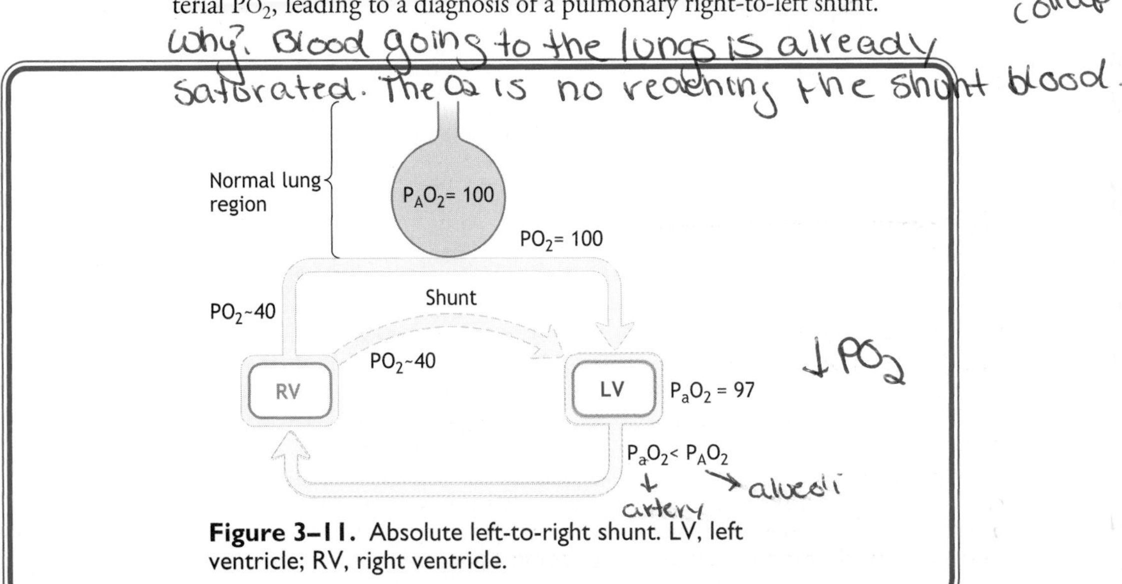

Figure 3–11. Absolute left-to-right shunt. LV, left ventricle; RV, right ventricle.

Thus, **overventilating** part of the lung does not compensate for the shunt because the empty Hb-binding sites in the shunted blood will bind the dissolved O_2 from the ventilated part of the lung, only slightly increasing PO_2 levels.

d. A **physiologic shunt** is the amount of absolute shunt that would cause the observed A-a difference.

2. In a **left-to-right shunt,** the pressures are higher in the left side of the heart; therefore, **hypoxemia is absent.** This type of shunt can be **due to arterial or ventricular septal defects or patent ductus arteriosis** (Table 3–2).

ATELECTASIS

- *Atelectasis is the collapse of the alveoli.*
- *Fever within 24–48 hours of surgery is usually due to atelectasis.*
- *Atelectasis causes pulmonary shunt (perfusion with no ventilation) and increased A-a gradient with hypoxemia.*
- *Hypoxemia is due to*
  - *–Right-to-left shunt*
  - *–Alveolar hypoventilation*
  - *–Diffusion abnormalities*

X. Ventilation-Perfusion Differences (Figure 3–12)

A. The relative difference between alveolar ventilation (V_A) and blood flow (Q)

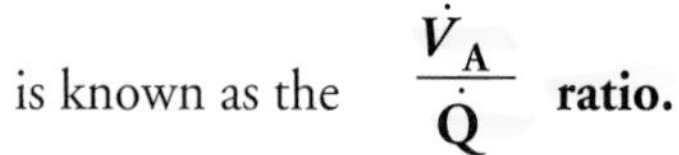

is known as the $\dfrac{\dot{V}_A}{\dot{Q}}$ **ratio.**

Table 3–2. Consequences of three different left-to-right shunts.

Measure	Atrial Septal Defect[a]	Ventricular Septal Defect[b]	Patent Ductus (Newborn)[c]
Systemic arterial Po_2	No change	No change	No change
Right atrial Po_2	⇑	No change	No change
Right ventricular Po_2	⇑	⇑	No change
Pulmonary arterial Po_2	⇑	⇑	⇑
Pulmonary blood flow	⇑	⇑	⇑
Pulmonary arterial pressure	⇑	⇑	⇑

[a]Atrial septal defect: Po_2 increase first appears in the right atrium.
[b]Ventricular septal defect: Po_2 increase first appears in the right ventricle.
[c]Patent ductus: Po_2 increase appears in pulmonary artery.

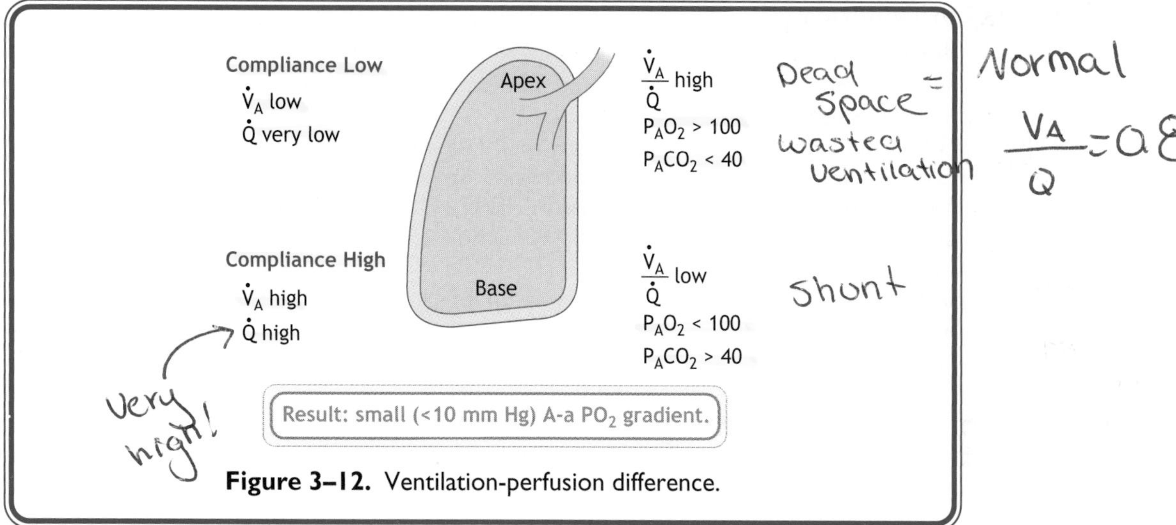

Figure 3–12. Ventilation-perfusion difference.

B. Thus, the **local alveolar gas composition** is not determined by ventilation alone or by blood flow (ie, perfusion) alone but by the ratio between ventilation and perfusion. In the normal lung, the ratio is approximately 0.8.

C. **Physiologic dead space** is defined as anatomic dead space plus the volume of all airways that behave as if they have received no blood flow.
  1. In health, anatomic dead space and physiologic dead space are essentially equal.
  2. In **ventilation-perfusion mismatch,** the amount of physiologic dead space is much greater than the amount of anatomic dead space.
     a. Some regions of the lung may have **a high CO_2**, which causes Po_2 in these alveoli **to be below average.**
     b. The **Bohr method** measures the volume of all airways in which no CO_2 has been added from the blood; this is the physiologic dead space.
     c. In many pulmonary diseases, the physiologic shunt and the physiologic dead space will be increased.
     d. The consequence of increased physiologic dead space is wasted ventilation.

D. **Hypoventilation is associated with equal decreases in Po_2** in the alveolar, pulmonary end capillary and systemic arterial compartments. **Supplemental oxygen or increased alveolar ventilation will return arterial Po_2** to normal.

E. **Diffusion impairment** refers to a lung structural problem (eg, increased thickness of lung membrane).
  1. With significant diffusion impairment, the **A-a gradient** widens.
  2. **Supplemental oxygen** will increase the gradient across the alveolar membranes and **return arterial PO_2 toward normal.**

Important: this is how you tell the difference between block + shunt!

F. **Exercise** increases ventilation and pulmonary blood flow. During exercise, the alveolar ratio is greater than 0.8, ventilation increases more than cardiac output, and base-to-apex flows become more equal.

XI. Special Environments

A. High Altitude

1. **At high altitude,** barometric pressure and ambient PO_2 decrease proportionally, **resulting in hypoxemia.**
2. Low PO_2 stimulates peripheral chemoreceptors, inducing hyperventilation, a decrease in alveolar and arterial PCO_2, and respiratory alkalosis.
3. Local **hypoxemia stimulates erythropoietin (EPO)**, a hormone produced by the kidney that increases red blood cell production and **can lead to polycythemia.** The increased Hb production increases the O_2 content of the blood.
    a. Because the kidneys are the major source of EPO, renal failure leads to anemia.
4. 2,3-DPG levels increase, shifting the oxyhemoglobin dissociation curve to the right and facilitating O_2 extraction by the tissues.
5. Hypoxemia also **results in hypoxic vasoconstriction** (ie, pulmonary vasoconstriction), resulting eventually in hypertrophy of the right ventricle due to increased work of the right heart.

B. Hyperbaric Chamber

1. Breathing room air (21% O_2; 79% N_2) in a hyperbaric environment increases the partial pressure of O_2 and N_2 in alveoli and arterial blood. Elevated PO_2 can produce oxygen toxicity, and the high PN_2 can lead to the **bends** (also known as caisson disease).
2. Sudden decompression causes bubbles of nitrogen to accumulate in the blood and tissues. Treatment is recompression and gradual decompression.

CLINICAL PROBLEMS

A 60-year-old man comes to your office complaining of dyspnea (difficult breathing). His laboratory values were as follows:

	Breathing Air	**Breathing 100% O_2 for 7 Minutes**
Arterial PO_2 (mm Hg)	76	500
Arterial PO_2 (mm Hg)	55	65
Arterial Hb (g/100 mL)	18	18
Arterial O_2 Sat (%)	85	100
Arterial pH	7.35	7.20

1. Which of the following diagnoses best explains the laboratory test findings?
   A. A patent ductus arteriosus
   B. A patent foramen ovale
   C. Complete obstruction of the right bronchus
   D. Thickened alveolus membrane impairing diffusion
   E. Pulmonary fibrosis restricting lung movement

Figure 3–13 shows the oxyhemoglobin dissociation curves for a healthy patient and for an anemic patient.

2. Which of the following statements concerning these patients is true?
   A. Patient A is anemic.
   B. Arterial PO_2 is likely to be similar for both subjects.
   C. Venous PO_2 of the anemic subject will be greater than that of the normal subject at rest or during exercise.
   D. If cardiac output is identical, then oxygen delivery will be identical in subjects A and B.

A 25-year-old, 70-kg man broke several ribs as a result of a fall from a ladder. His treatment at a nearby hospital included stabilizing his chest with bandages. The bandages were tied in a way that reduced his tidal volume by 50%. To compensate, he doubled his respiratory rate. Two hours later, an arterial blood sample was taken.

3. Which of the following conditions would have been observed?
   A. Increased PO_2 and decreased PCO_2
   B. No change in PO_2 or PCO_2

$V_A = (V_T - V_D) \times$ Breaths/min

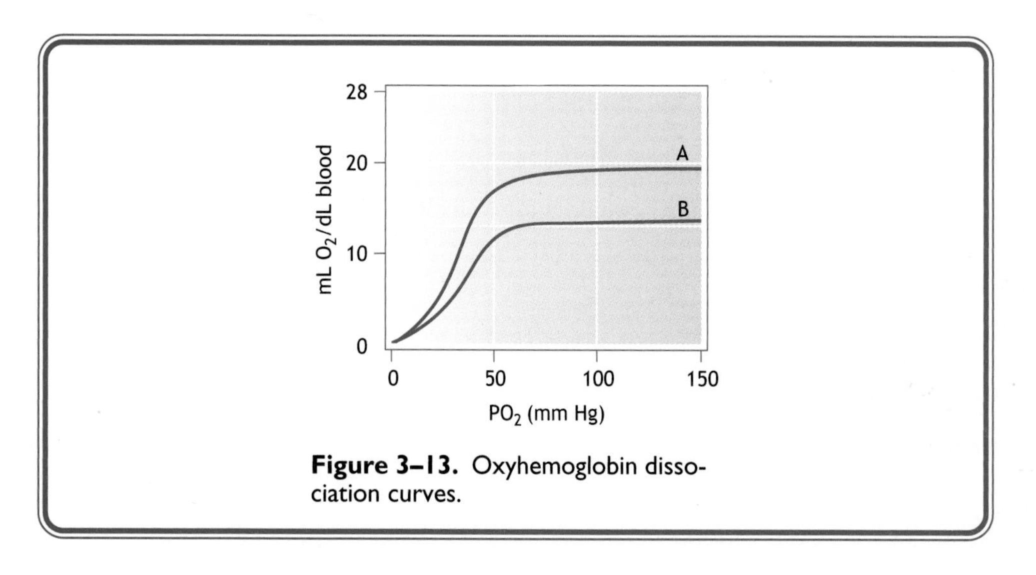

Figure 3–13. Oxyhemoglobin dissociation curves.

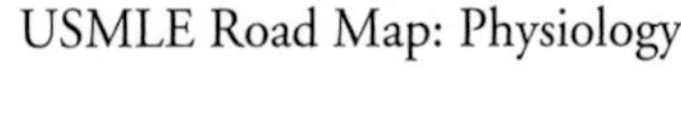

C. Decreased PO_2 and increased PCO_2

D. Increased PO_2 and increased PCO_2

A patient has an increased airway resistance to gas flow but a normal compliance.

4. In comparison to the findings in a healthy person, intrapleural pressure will be
   A. More positive during inspiration
   B. More negative during expiration
   C. Increased at functional residual capacity
   D. Normal during breath holding at total lung capacity
   E. Decreased during breath holding at total lung capacity

An individual's total lung capacity (TLC) is 6.5 L, and her inspiratory capacity (FRC – TLC) is 3.55 L. At the end of a normal expiration, her lung volume is 4.45 L.

5. The individual's tidal volume (V_T) is
   A. 1.50 L
   B. 3.00 L
   C. 0.500 L
   D. 0.750 L
   E. 0.900 L

Following infusion of lactic acid into the blood of a healthy subject, arterial pH falls to 7.35.

6. Which of the following would be expected to occur?
   A. A decrease in ventilation
   B. A rise in the pH of the cerebrospinal fluid
   C. A decrease in arterial PO_2
   D. A rise in arterial PCO_2
   E. A decrease in the ratio

7. Which of the following causes of brain hypoxia would most strongly stimulate the aortic and carotid chemoreceptors?
   A. Carbon monoxide poisoning
   B. Severe anemia
   C. Formation of methemoglobin
   D. A marked decrease in pulmonary diffusing capacity
   E. Acute respiratory alkalosis

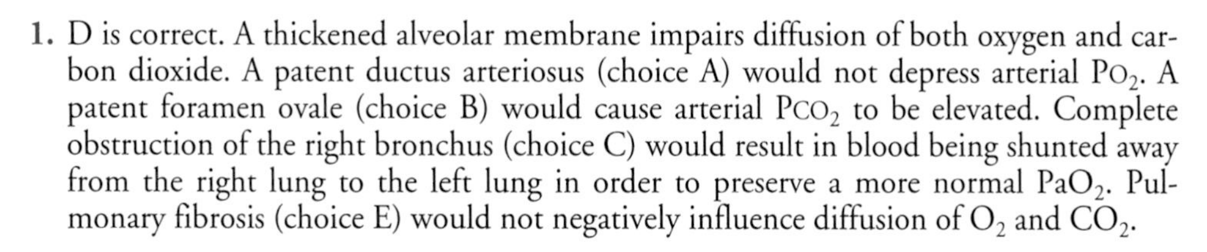

ANSWERS

1. D is correct. A thickened alveolar membrane impairs diffusion of both oxygen and carbon dioxide. A patent ductus arteriosus (choice A) would not depress arterial PO_2. A patent foramen ovale (choice B) would cause arterial PCO_2 to be elevated. Complete obstruction of the right bronchus (choice C) would result in blood being shunted away from the right lung to the left lung in order to preserve a more normal PaO_2. Pulmonary fibrosis (choice E) would not negatively influence diffusion of O_2 and CO_2.

2. B is correct. Arterial PO_2 would be similar for both because the PO_2 is indicative of dissolved plasma oxygen, not oxygen combined with hemoglobin. Choice A is incorrect because the O_2 in the blood is decreased in patient B, not patient A. Choice C is incorrect because PO_2 is indicative of dissolved oxygen, which should be the same in both individuals at rest or during exercise. Choice D is incorrect because oxygen delivery to the tissues depends on both oxygen bound to hemoglobin and dissolved oxygen. In the anemic individual, less oxygen is delivered because less hemoglobin is bound to oxygen.

3. C is correct. A reduction in alveolar ventilation (inspired air available for exchange) would decrease PO_2 and increase PCO_2. Choices A and B are incorrect because arterial blood samples would show a decrease in O_2 and an increase in CO_2. Choice D is incorrect because there would be no increase in oxygen with a reduction in alveolar ventilation.

4. D is correct. During breath holding at total lung capacity, the intrapleural pressure would be normal when no air is moving. Choice A is incorrect because during inspiration, intrapleural pressure is more negative due to increased airway resistance. Choice B is incorrect because positive elastic recoil during expiration makes intrapleural pressure more positive. Choice C is incorrect because functional reserve capacity is the volume of air remaining in the lungs after a normal expiration and no air is moving; thus, intrapleural pressure is not increased. Choice E is incorrect because intrapleural pressure will be normal when no air is moving.

5. A is correct. It is calculated as follows: The individual's inspiratory capacity is 3.55 L, and the volume at the end of a normal expiration is 4.45 L. Because these volumes contain the tidal volume, they must be summed (3.55 + 4.45 = 8.0 L). The TLC (6.5 L) is then subtracted from 8.0 L. Therefore, 8.0 L – 6.5 L = 1.5 L, the individual's tidal volume.

6. B is correct. Infusion of lactic acid will decrease the partial pressure of CO_2 in the blood and cause diffusion of CO_2 from the cerebrospinal fluid to the blood, thereby increasing the pH of the cerebrospinal fluid. A decrease in ventilation (choice A) would not occur because acidosis would increase, not decrease, ventilation. A decrease in arterial PO_2 (choice C) would not occur because ventilation would be increased, enhancing arterial PO_2 levels. A rise in arterial PCO_2 (choice D) would not occur because lactic acid infusion would reduce the arterial CO_2. A decrease in the $\frac{\dot{V}_A}{\dot{Q}}$ ratio (choice E) would not occur because ventilation would be increased, not decreased.

7. D is correct. A marked decrease in pulmonary diffusing capacity decreases PaO_2 and increases $PaCO_2$, both of which would increase the firing of peripheral chemoreceptors. Choices A and B are incorrect because carbon monoxide poisoning and severe anemia reflect less O_2 binding to Hb, but this does not alter PaO_2 because it reflects dissolved O_2, not O_2 combined with Hb. Formation of methemoglobin (choice C) occurs when the ferrous iron of the heme molecule is converted to ferric iron, but this choice is incorrect because hemoglobin binding has little to do with stimulation of peripheral chemoreceptors by dissolved O_2. Respiratory alkalosis (choice E) would not stimulate peripheral chemoreceptors because $PaCO_2$ levels are decreased, not increased.

CHAPTER 4

BODY FLUIDS, RENAL, AND ACID-BASE PHYSIOLOGY

I. Body Fluids

A. Humans are composed primarily of water.

B. Body composition depends on age and sex (Table 4–1).

C. Total body water (**TBW**) is divided into two major compartments: the intracellular fluid (**ICF**) and the extracellular fluid (**ECF**) (Figure 4–1).

1. The ICF compartment represents fluid contained within all the cells in the body, or approximately two-thirds of TBW.
2. The ECF compartment includes all fluids outside of cells and represents approximately one-third of TBW. ECF is further divided into
    a. Blood plasma (blood without cells)
    b. Interstitial fluid (**ISF**) (fluid between cells)
    c. Transcellular fluid (synovial intraocular, pericardial, cerebrospinal, and epithelial fluids)

D. Adipose tissue is low in water content; thus, obese individuals have a lower fraction of body weight that is water than do normal-weight individuals.

E. The measurement of body fluid compartments follows several principles.

1. Substances used to determine the volume of body fluid compartments must have the following characteristics:
    a. They must be nontoxic.
    b. They must not be synthesized or metabolized in the compartment measured.
    c. They must not induce shifts in fluid distribution among different compartments.
    d. They must be easily and accurately measured.
2. According to the **indicator dilution principle,** the volume of a fluid compartment can be calculated by measuring the concentration of an indicator injected into the compartment.
3. The larger the volume of fluid in which the substance is diluted, the more the substance is diluted.
4. A known volume (V_1) of an indicator is injected into the body. The quantity of the indicator (Q) injected equals its concentration (C_1) times its volume (V_1). After equilibration in the body fluid compartment, the concentration will become C_2; thus, $Q = C_2V_2$. Therefore, the volume of the body fluid compartment (V_2) is calculated as follows:

Table 4–1. Water percentage of body weight based on age and gender.

Age	Male	Female
Newborn	80%	75%
1–5 years	65%	65%
10–16 years	60%	60%
17–39 years	60%	50%
40–59 years	55%	47%
60+	50%	45%

$$V_2 = \frac{amount\ (Q)}{concentration\ (C_2)}$$

a. For example, plasma volume can be measured using radioiodinated serum albumin (^{131}I-albumin, or **RISA**) or **Evans blue dye.**
b. 350,000 counts per minute (cpm) of RISA is injected intravenously (Q = 350,000 cpm). After a 1-hour equilibration period, 10 mL of blood is withdrawn and centrifuged to separate the blood cells from the plasma, and 1 mL of plasma is found to contain 100 cpm.
c. If $Q = C_2V_2$, then 350,000 cpm = 100,000/L × V_2, or plasma volume $V_2 = Q/C_2$, or 3.5 L.

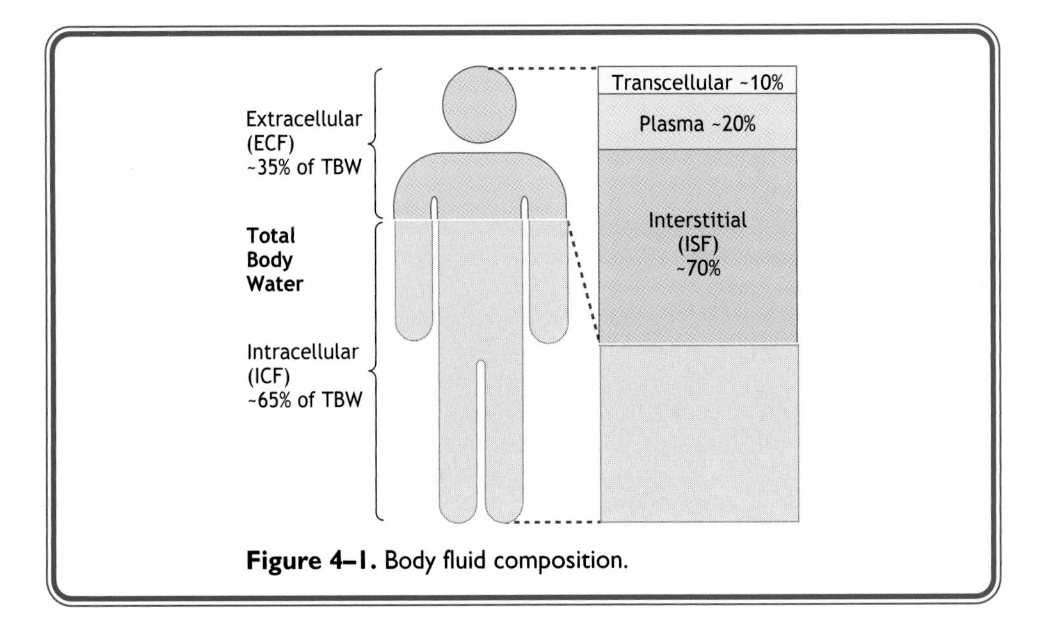

Figure 4–1. Body fluid composition.

5. A variety of substances are used to measure major body fluid compartments.
   a. **Tritiated water** and **deuterium oxide** are used to measure **TBW.**
   b. **Inulin, mannitol,** and **sulfate** are used to measure **ECF.**
   c. **RISA** and **Evans blue** are used to measure **plasma.**
   d. ISF and ICF can then be calculated by using the following equations:

$$\text{ECF} - \text{plasma volume} = \text{ISF}$$
$$\text{TBW} - \text{ECF} = \text{ICF}$$

6. Estimates of body fluid compartments can also be made at the patient's bedside (Table 4–2).

F. TBW daily turnover due to water intake and loss is shown in Figure 4–2.
1. **Water intake** averages about 2 L/d, although this amount is highly variable.
2. **Insensible water loss** is approximately 0.74 L/d due to water evaporation through the skin and due to respiration.
   a. **Water loss through the skin** (**0.3–0.4 L/d**) is not dependent on sweating and occurs in people born with no sweat glands. The rate of water loss is minimized because of the cornified layer of the skin. When this skin layer is lost following severe burns, the rate of water loss increases dramatically to about 3–5 L/d.
   b. **Water loss due to respiration is about 0.3–0.4 L/d.** Water vapor pressure in the lung is approximately 47 mm Hg. Inspired air becomes saturated with moisture because it has a lower vapor pressure. In cold weather, vapor pressure in the air decreases even further, enhancing water loss.
3. At rest, **water loss due to sweating is approximately 0.1 L/d,** but this amount increases dramatically during heavy exercise (up to 1–2 L/h).
4. **Feces accounts for approximately 0.1–0.2 L/d.** This amount increases dramatically with diarrhea.

Table 4–2. Bedside estimates of body fluid compartment volumes.

Remember	Example for 60-kg Patient
Total body water = 60% × body weight	60% × 60 kg = 36 L
Intracellular fluid = 2/3 total body water	2/3 × 36 L = 24 L
Extracellular fluid = 1/3 total body water	1/3 × 36 L = 12 L
Plasma volume = 1/4 extracellular water	1/4 × 12 L = 3 L
Blood volume = $\frac{\text{Plasma volume}}{1 - \text{Hct}}$	3 L ÷ (1 − 0.40) = 6.6 L

Hct, hematocrit

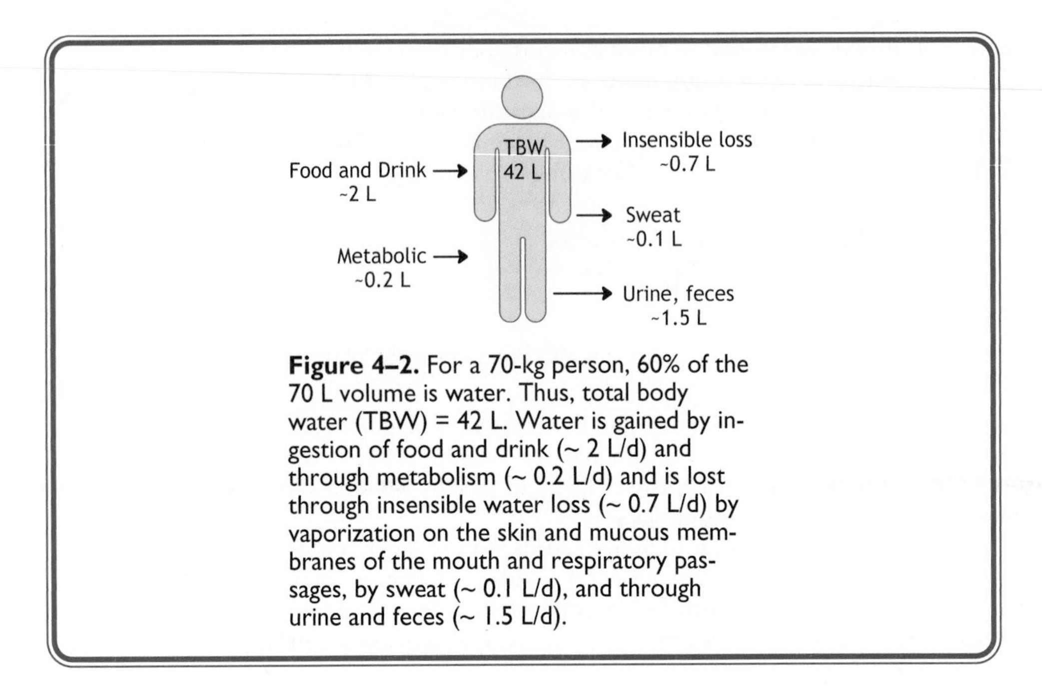

Figure 4–2. For a 70-kg person, 60% of the 70 L volume is water. Thus, total body water (TBW) = 42 L. Water is gained by ingestion of food and drink (~ 2 L/d) and through metabolism (~ 0.2 L/d) and is lost through insensible water loss (~ 0.7 L/d) by vaporization on the skin and mucous membranes of the mouth and respiratory passages, by sweat (~ 0.1 L/d), and through urine and feces (~ 1.5 L/d).

5. **Urine accounts for approximately 0.5–1.5 L/d** but varies depending on the level of water intake. Water excretion through the kidneys constitutes the major regulator of body water and electrolyte balance.

G. Fluid shifts between compartments follow several basic principles:
1. ICF and ECF are in osmotic equilibrium.
2. **Na^+** is the major cation of the **ECF.**
3. **K^+** is the major cation of the **ICF.**
4. The distribution of Na^+ and K^+ is maintained by Na^+-K^+-ATPase.
5. Equilibration between the ICF and ECF occurs through **water movement,** not through movement of osmotically active particles.
6. **Fluids move** unassisted across cell membranes **only because of osmolarity differences.**
7. Table 4–3 illustrates the effect of various conditions on TBW, ECF, ICF, ECF osmolarity, and serum Na^+ levels.

SODIUM DISORDERS

- ***Serum osmolality*** *is the amount of solute concentration in a solution.*
- ***Sodium*** *is the key component that determines serum osmolality.*
- *The* ***ECF sodium concentration*** *is responsible for the movement of water between the ECF and ICF compartments.*
- *Either* ***hyponatremia*** *(decrease in serum Na^+) or* ***hypernatremia*** *(increase in serum Na^+) produces an osmotic gradient between the ECF and ICF.*
  - ***Hyponatremia causes water to move into the ICF.***
  - ***Hypernatremia causes water to move out of the ICF.***
  - *Glucose is limited to the ECF compartment.*

Table 4–3. Effects of various conditions on body fluid composition.

Condition	TBW	ECF	ICF	ECF Osmolarity	Serum Na
Isotonic NaCl infusion	↑	↑	NC	NC	↔ Na
Diarrhea	↓	↓	NC	NC	↔ Na
Excessive NaCl intake	↑	↑	↓	↑	↑ Na
Excessive sweating	↓	↓	↓	↑	↑ Na
Excessive ADH (SIADH)	↑	↑	↓	↓	↓ Na
Adrenal insufficiency	↓	↓	↑	↓	↓ Na

ADH, antidiuretic hormone;
NC, no change.

*–**Hyperglycemia** (increased serum glucose) causes water to move out of the ICF, overriding the role of sodium as the major osmotic force.*

- ***Tonicity of plasma** refers to factors (ie, Na^+ and glucose) that cause water movement between the ECF and ICF; it is not the same as osmolality.*

II. Kidney Function

A. The kidneys have **excretory, endocrine,** and **regulatory functions.**

B. The kidneys regulate the composition and volume of body fluids by excreting or conserving the correct amounts of water and solutes.

C. The kidneys act as endocrine organs, releasing **renin, erythropoietin,** and **1,25-dihydroxy-vitamin D_3** into the circulation.

D. The kidneys excrete metabolic end-products (ie, **urea, uric acid,** and **creatinine**) and foreign substances.

E. Renal function is based on four steps:

1. Blood from renal arteries is delivered to the glomeruli. At **one-fifth of cardiac output,** this is the **highest tissue-specific blood flow.**
2. Glomeruli form ultrafiltrate, which flows into renal tubules.
3. Tubules reabsorb and secrete solute and water from the ultrafiltrate.
4. Tubular fluid leaves the kidney via the ureter to the bladder and out through the urethra.

F. The amount of a substance excreted by the kidney is determined by the amount filtered by the glomerulus less the amount absorbed plus the amount secreted (Figure 4–3).

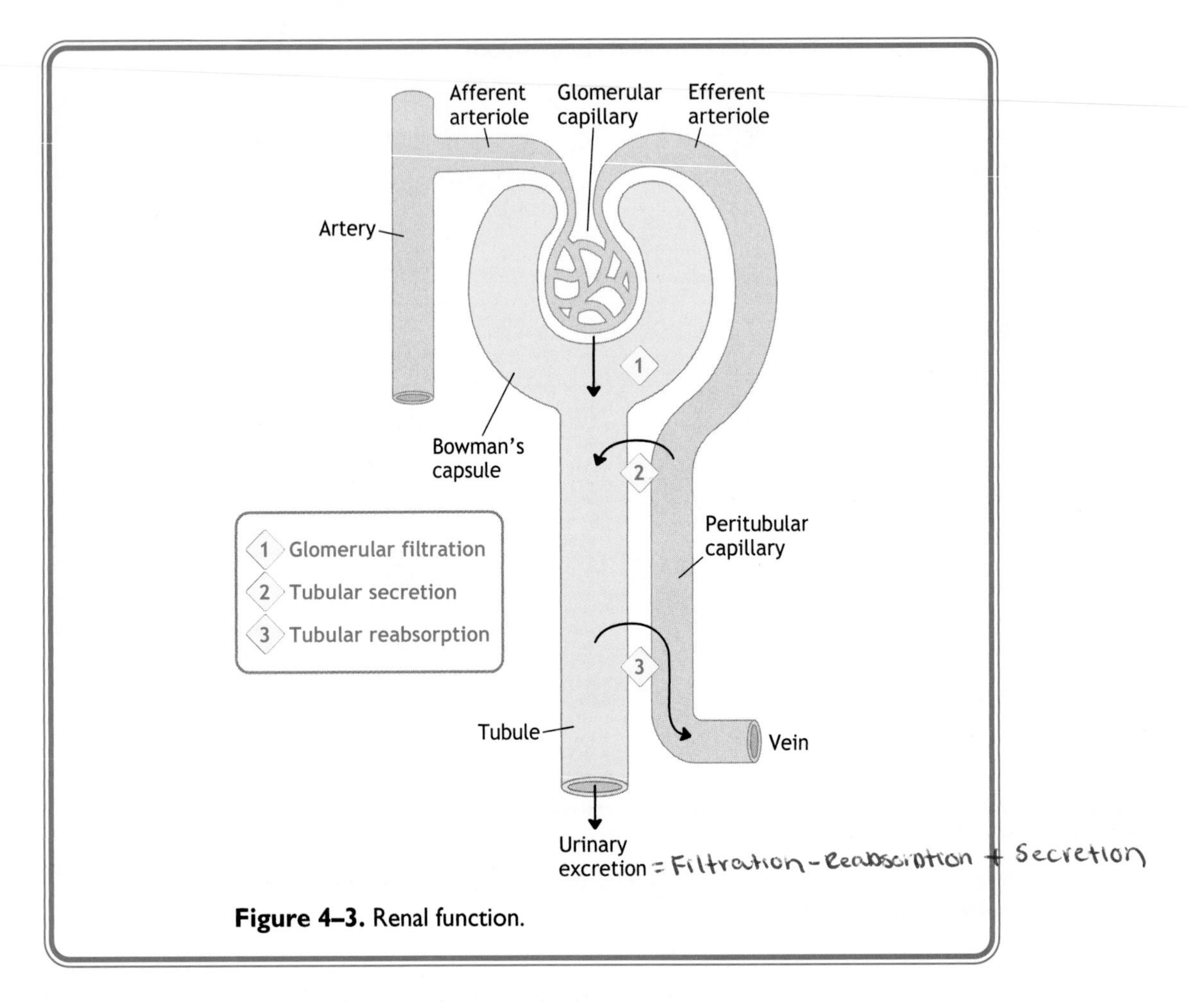

Figure 4–3. Renal function.

III. Renal Anatomy

A. The kidneys are paired, retroperitoneal organs that are composed of a mixture of vascular and epithelial elements.

B. The human kidney is multilobed and grossly divided into a cortex and a medulla.

C. The basic functional unit of the kidney is the **nephron** (Figure 4–4).

1. The nephron is composed of a long, thin tubule that is closed at one end (**Bowman's capsule**).
2. Bowman's capsule surrounds a **high-pressure capillary network,** the **glomerulus.**
3. Together, Bowman's capsule and the glomerulus serve as a filtration unit, which forms the glomerular filtrate that enters the tubule.
4. Tubular components of the nephron include the proximal tubule, Loop of Henle, distal tubule, and collecting duct.
   a. The tubular fluid generated by glomerular filtration is modified by reabsorption and secretion across the epithelial cells that form the tubule wall.

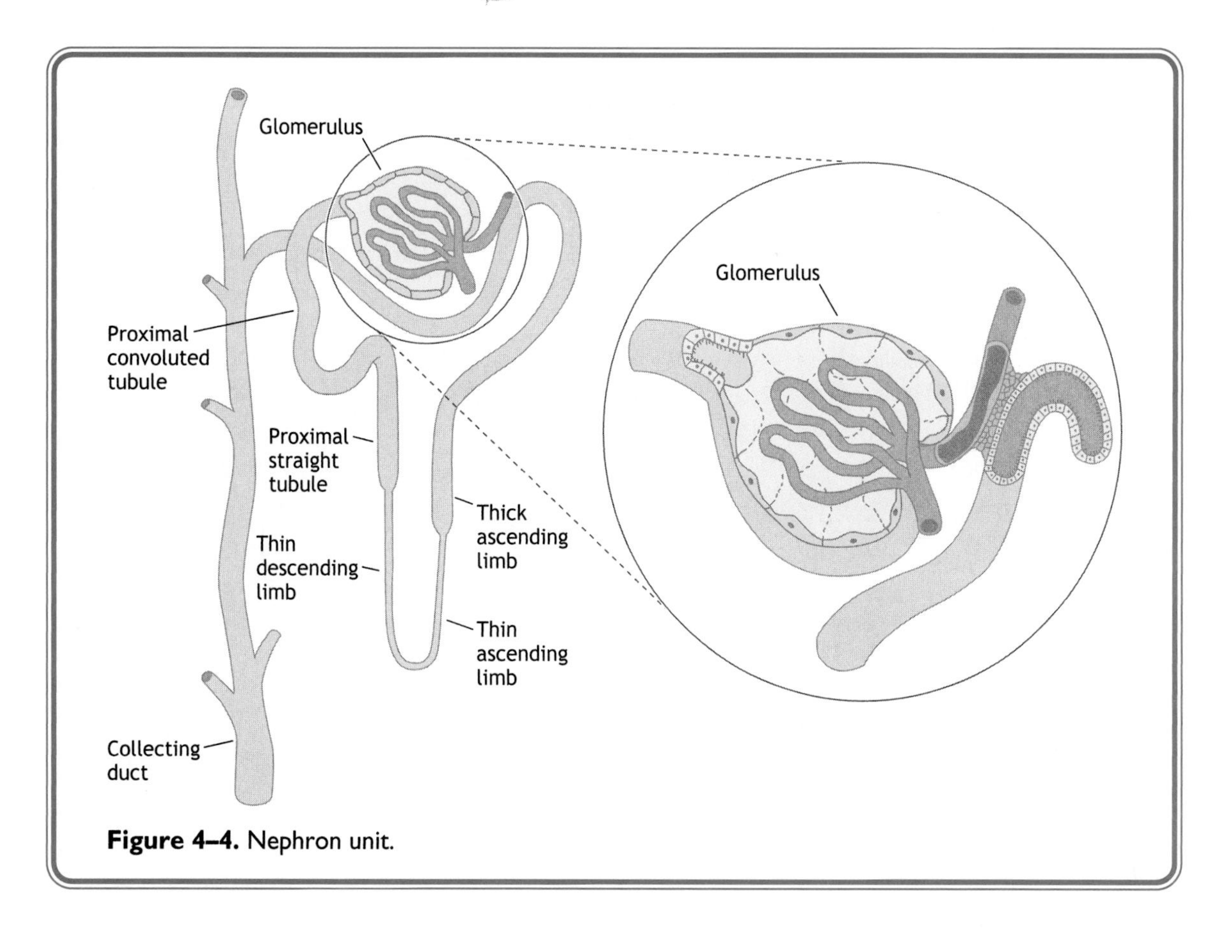

Figure 4–4. Nephron unit.

   b. Net movement of water or solutes from the tubular lumen into the interstitium is referred to as tubular **reabsorption.**
   c. Net transport of substances from the interstitium to the lumen is called **secretion.**
   d. **Excretion** refers to removal of a substance from the body.
      (1) Renal excretion of water is referred to as **diuresis.**
      (2) Renal excretion of sodium is referred to as **natriuresis.**

5. Each nephron has its own blood supply, which is composed of two arterioles and two capillary systems in series (Figure 4–5).
   a. The first capillary system is a high-pressure capillary **glomerulus** that favors filtration and is the source of the tubular fluid.
   b. After passing through the **afferent arteriole,** the **glomerulus,** and the **efferent arteriole,** the blood enters the **peritubular capillary system,** a low-pressure system that favors reabsorption.
   c. Sympathetic nerve fibers regulate blood flow, glomerular filtration, and tubule reabsorption.

D. There are two types of nephrons: **cortical** (about 85%) and **juxtamedullary** (15%).

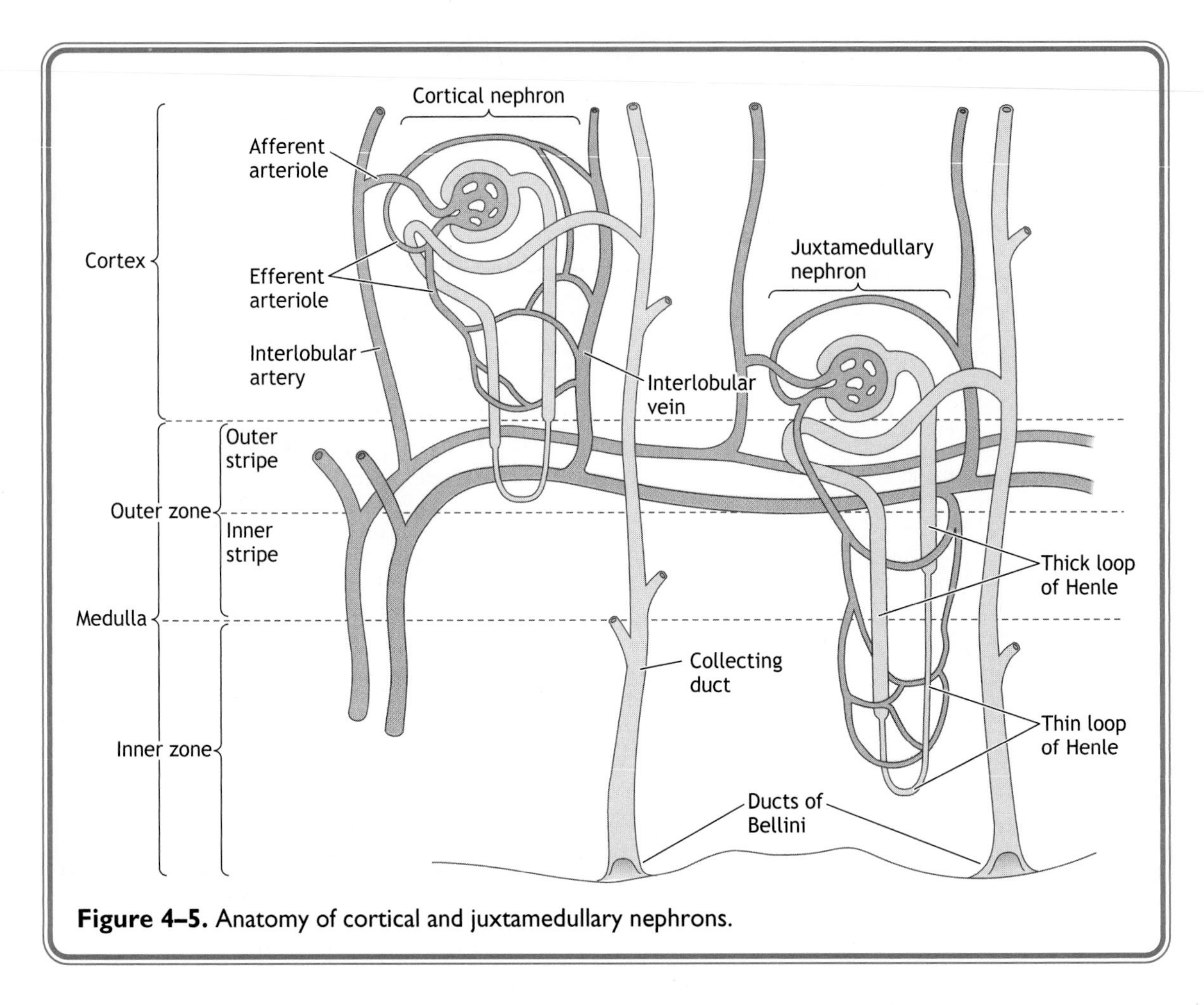

Figure 4–5. Anatomy of cortical and juxtamedullary nephrons.

1. Cortical nephrons have short loops of Henle with peritubular capillaries. Peritubular capillaries differ depending on their association with different nephrons.
2. Juxtamedullary nephrons have long loops of Henle and vasa recta (see Figure 4–5). The **vasa recta** are long narrow capillary tubules that have a great resistance to blood flow.

E. A portion of the arterial plasma leaves the glomerulus (ie, the product of **filtration**) to form the protein-free tubular fluid. The remaining arterial plasma enters the peritubular capillary system or vasa recta.

F. The kidneys are also endocrine organs producing renin, erythropoietin, prostaglandins, 1-25 OH^2 vitamin D, and bradykinin.

ADULT POLYCYSTIC KIDNEY DISEASE

- *Adult polycystic kidney disease is the* ***most common inherited disorder*** *of the kidney.*
- *The disease is autosomal dominant and is characterized by slow progression. In 90% of cases, the locus for the disease is on the short arm of chromosome 16.*

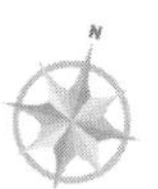

- *A highly polymorphic locus is associated with the α-globin gene cluster, which must be coinherited with the disease.*
- *Symptoms include lumbar back pain, **hematuria** (blood in urine), infection, hypertension and **nephrolithiasis** (kidney stones), and renal failure.*
- *The external surface of the kidney shows multiple cysts.*

IV. Renal Blood Flow and Glomerular Filtration

A. Approximately 25% of cardiac output supplies the kidneys, which account for about 1% of body mass.

B. The high blood flow to the kidney is necessary to generate the large hydrostatic pressure responsible for the formation of glomerular ultrafiltrate.

C. Most of the renal blood flow goes to the cortex, where the glomeruli are located.

D. Renal blood flow remains constant over a wide range of arterial pressures. This **autoregulation** is accomplished by increases in afferent arteriolar resistance.

E. **Renal handling of *p*-aminohippuric acid (PAH)** is an example of active secretion by a **transport-maximum (Tm)-limited mechanism.**

1. **PAH** is foreign to the body and is excreted by filtration plus secretion.
2. The active secretory mechanism is located on the basolateral membrane of the proximal convoluted tubule.
3. The PAH carrier is saturable and **is inhibited by** the drug **probenecid.**
4. The transport of PAH increases linearly with the concentration of PAH (P_{PAH}) until the delivery of PAH to the peritubular capillaries increases to the point where Tm is attained. Secretion of PAH then becomes constant (Tm_{PAH}) and equals about 80 mg/min/1.73 m^2 in a young male adult.
5. Because PAH is actively secreted in the proximal tubular segment, the **Tm_{PAH} is a measure of the functional mass of proximal tubules.**

Renal plasma flow

F. The rate at which tubular fluid is produced is termed the **glomerular filtration rate** (GFR) (normally 120–125 mL/min).

1. The driving force for glomerular filtration is the net **ultrafiltration pressure,** which always favors fluid movement out of the capillaries.
2. The glomerulus is a high-pressure capillary system, and the peritubular capillary system (as well as the vasa recta) is a low-pressure system. Thus, the GFR can be related to **Starling's forces:**

$$GFR = kS \times (P_{gc} + \pi_t - P_t - \pi_{gc}),$$

where
P_{gc} = glomerular capillary hydrostatic pressure (mm Hg)
P_t = tubule lumen hydrostatic pressure (mm Hg)
π_{gc} = glomerular capillary colloid osmotic pressure (mm Hg)
π_t = tubule lumen colloid osmotic pressure (mm Hg)
kS = coefficient relating net Starling's force to GFR
GFR = globular filtration rate (measured in mL/min; varies with surface area, which is 1.73 m^2)

3. In the glomerular capillary, hydrostatic pressures (P_{gc}) provide the driving force for filtration, whereas the oncotic pressures in the capillary and the hydrostatic pressure in Bowman space oppose it.

4. Because tubular filtrate is virtually protein free, proteins are concentrated in the glomerulus, and colloid osmotic pressure (π_{gc}) increases as blood flows through the glomerulus. As π_{gc} rises and meets the hydrostatic pressure, **filtration equilibrium** (ie, where net filtration pressure is 0) is attained.
5. An **increase in blood flow** through the glomerulus **increases the GFR** because it increases the distance over which filtration occurs before equilibrium is reached.
6. Thus, variables that influence GFR include
   a. **Hydrostatic pressures in the glomerulus and Bowman's capsule**
   b. **Oncotic pressures in the glomerular plasma and filtrate**
   c. **Permeability of glomerular barriers**
   d. **Surface area available for filtration**
   e. **Negative electrical charge on filtered solutes** (which hinders filtration)
   f. **Renal blood flow**

NEPHROTIC SYNDROME

- *Nephrotic syndrome is a glomerular disease* ***characterized by proteinuria, edema, lipiduria, hypercoagulation,*** *and* ***hyperlipidemia.***
- *Proteinuria is due to decreased charge or size selectivity by the glomerulus or increased permeability of the glomerulus.*
- *Classic physical finding are horizontal white bands on the fingernails.*

NEPHRITIC SYNDROME

- *Nephritic syndrome is the* ***result of diffuse glomerular inflammation.***
- *The syndrome is* ***characterized by*** *1) sudden onset of* ***hematuria,*** *2)* ***decreased GFR*** *resulting in increased BUN (blood urea nitrogen) and creatinine, 3)* ***oliguria,*** *4)* ***hypertension,*** *and 5)* ***edema.***

7. **Clearance** is defined as the volume of arterial plasma that would be totally cleared of a solute in a given time. Understanding the concept of clearance is critical to evaluating renal function. Symbolically,

$$C_x = \frac{U_x \times V}{P_x},$$

where
C_x = clearance of x (mL/min)
U_x = urine concentration of x (mg/mL)
P_x = arterial plasma concentration of x (mg/mL)
V = urine flow (mL/min)
   a. Clearance is usually measured in the steady state. If a **substance is present in arterial plasma but is not excreted** (U_x = 0), then the **clearance of that substance is 0.**
   b. If a substance is secreted into the tubular fluid (eg, a foreign substance), then it cannot be excreted at a rate faster than it is presented to the kidneys via the renal arteries (ie, clearance cannot exceed renal plasma flow).
   c. **If a substance X is freely filterable and is not secreted or reabsorbed** by the tubules, **then the clearance of X can be used to measure the GFR.** Thus, the GFR for substance X is identical to its clearance:

$$GFR = \frac{U_X \times V}{P_X} = C_X$$

8. The clearance of inulin is a measure of GFR.
   a. Inulin is a fructose extracted from dahlia roots.

9. Inulin meets the criteria for measuring GFR by the clearance method.
   a. H is **freely filtered** (ie, not bound to plasma protein).
   b. It is **not reabsorbed or secreted.**
   c. It is **not stored or metabolized by the kidneys.**
   d. It is **nontoxic.**
   e. It does **not alter the GFR.**
   f. It should be **easily measurable in plasma and urine.**

10. The clearance of creatinine is a useful clinical index of GFR. Because creatinine is endogenous, it does not need to be infused and is used clinically to assess glomerular function.
   a. Although C_{cr} provides an estimate of GFR and glomerular function, measurement of U_{cr} and urine volume are cumbersome; thus, **elevated P_{cr}** provides an indicator of reduced GFR.
   b. In general, C_{cr} overestimates the GFR by 15–20% **because the GFR decreases with age but P_{cr}** remains constant due to decreased muscle mass.
   c. Consider the following example of a GFR calculation: A 68-kg patient has a urine volume of 1.5 L/24 h. The U_{cr} is 0.9 mg/mL, and the P_{cr} is 0.8 mg/100 mL.

$$\text{Urine flow} = V = \frac{1500 \text{ mL/24h}}{1440 \text{ min/24h}} = 1.04 \text{ mL/min}$$

$$P_{cr} = \frac{0.8 \text{ mg}}{100 \text{ mL}} = 0.008 \text{ mg/mL}$$

$$GFR = \frac{U_{cr} \times V}{P_{cr}} = \frac{(0.9 \text{ mg/mL} \times 1.04 \text{ mL/min})}{0.008 \text{ mg/mL}} = 117 \text{ mL/min}$$

   d. **Filtration fraction (FF)** is the fraction of renal plasma volume (RPF) that is filtered at the glomerulus. Thus,

$$FF = \frac{GFR}{RPF}$$

$$RPF = RBF \times (1 - Hct),$$

where

$$RBF =\sim 1.25 \text{ L/min} = \frac{\textit{aortic pressure} - \textit{renal venous pressure}}{\textit{renal vascular resistance}}$$

Hct = hematocrit (%)

(1) Normally 20% of the RBF is filtered, and the remainder flows into the peritubular capillary.

(2) **An increase in FF causes an increased protein concentration in peritubular capillary blood.**

(3) **Increased postglomerular resistance increases FF,** and vice versa.

e. **Renal blood flow** (**RBF**) can be calculated from the RPF if the hematocrit (Hct, %) is known:

$$RBF = RPF \frac{100}{(100 - Hct)}$$

11. The GFR increases when glomerular capillary pressure is increased and decreases when glomerular capillary pressure is decreased.
12. Alterations in preglomerular and postglomerular renal vascular resistance influence RBF, GFR, and FF (Table 4–4).

G. **The clearance of para-aminohippurate (PAH) is a measure of renal plasma flow.**

1. Most of the arterial plasma entering the kidneys perfuses the proximal tubular segment.
2. Arterial plasma flow entering the kidneys splits into two parallel paths. **One path perfuses** the proximal tubular segment used for urine production (ie, **secretory tissue**), and **the other,** which keeps the tissue alive, **perfuses inert tissue.**
3. These facts are used to develop the **Fick method** for measuring total renal plasma flow using **PAH as the marker,** assuming that steady-state conditions exist (ie, PAH entering kidneys/min = PAH leaving kidneys/min):

$$Pa_{\text{PAH}}RPF = Pv_{\text{PAH}}RPF + U_{\text{PAH}}V$$
$$RPF\ (Pa_{\text{PAH}} - Pv_{\text{PAH}}) = U_{\text{PAH}}V$$

Thus,

$$RPF = \frac{U_{\text{PAH}}V}{(Pa_{\text{PAH}} - Pv_{\text{PAH}})},$$

Table 4–4. Consequences of independent isolated constrictions or dilations of the afferent and efferent arterioles.

Alteration	Glomerular Capillary Pressure	Peritubular Capillary Pressure	Nephron Plasma Flow
Constrict efferent	⇑	⇓	⇓
Dilate efferent	⇓	⇑	⇑
Constrict afferent	⇓	⇓	⇓
Dilate afferent	⇑	⇑	⇑

where
Pa_{PAH} = arterial plasma concentration of PAH
Pv_{PAH} = renal venous plasma concentration of PAH
RPF = total renal plasma flow

4. If it is assumed that the kidneys can remove all of the PAH from arterial plasma, then the venous plasma concentration of PAH would be 0, and the Fick equation would be equal to the clearance of PAH:

$$RPF = \frac{U_{PAH}V}{(Pa_{PAH} - 0)} = C_{PAH}$$

5. **Because a fraction of arterial plasma** does not perfuse nephrons but rather **perfuses inert tissue, the venous plasma concentration of PAH is not 0.** Pv_{PAH} concentration is about 10% of Pa_{PAH} when P_{PAH} levels are low.
6. Thus, if **C_{PAH} is measured when P_{PAH} levels** are low, **it is about 90% of total RPF** (Figure 4–6). C_{PAH} measures **effective renal plasma flow** (**ERPF**). Total renal plasma flow is designated as RPF.

V. Transport Mechanisms of Nephron Segments

A. Proximal Tubule (Figure 4–7)

1. **Loose** tight junctions make the proximal tubule water permeable.
2. The bulk of filtered small solutes is absorbed. For example, **60% of filtered Na^+, Cl^-, K^+, Ca^{2+},** and **H_2O** is absorbed; and **90% of filtered HCO_3-** is absorbed.

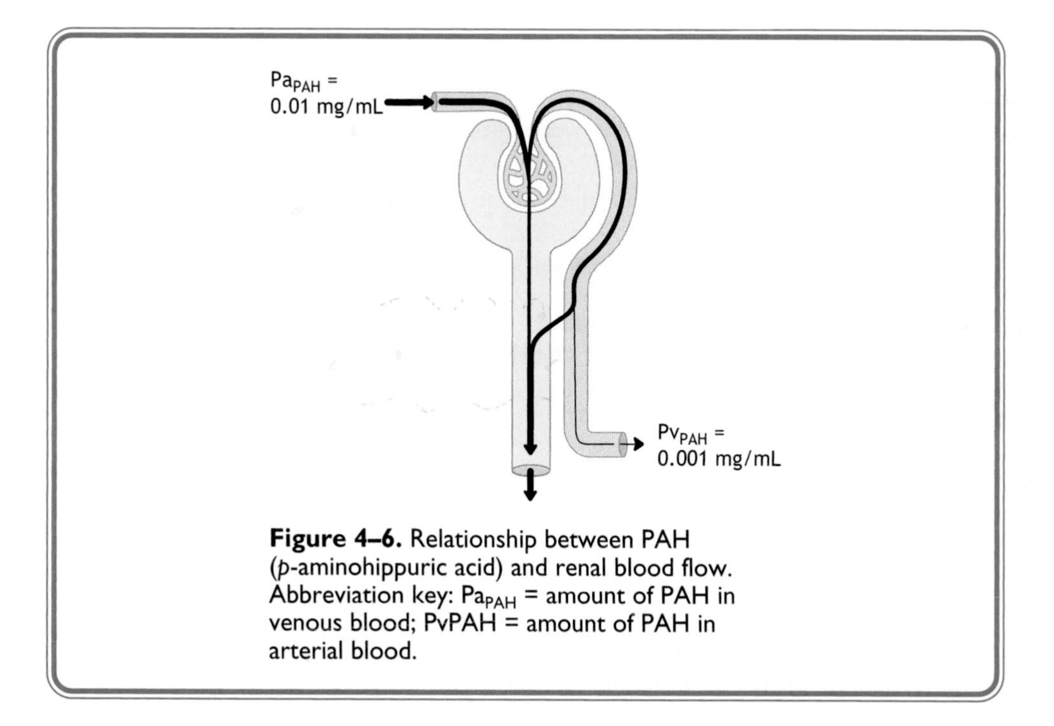

Figure 4–6. Relationship between PAH (*p*-aminohippuric acid) and renal blood flow. Abbreviation key: Pa_{PAH} = amount of PAH in venous blood; PvPAH = amount of PAH in arterial blood.

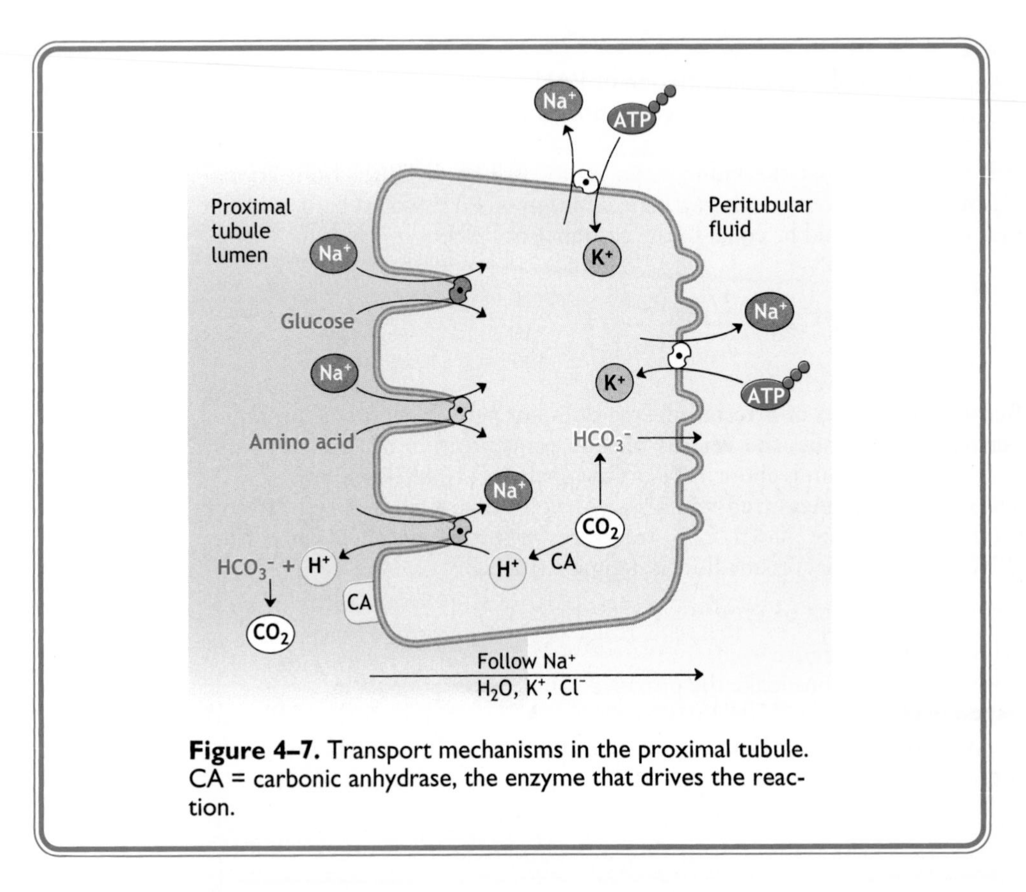

Figure 4–7. Transport mechanisms in the proximal tubule. CA = carbonic anhydrase, the enzyme that drives the reaction.

3. All filtered **glucose** and **amino acid is absorbed.**
4. **Phosphate transport** is regulated by parathyroid hormone.
5. Osmolarity does not change due to passive reabsorption of water.

B. Loop of Henle (Figure 4–8)

1. The volume of fluid reaching the loop of Henle is about one-third of the originally filtered volume.
2. The **descending limb is water permeable,** increasing the osmolarity of the tubular fluid.
3. The **ascending limb is impermeable to water,** decreasing the tubular fluid osmolarity. This segment is known as the **diluting segment** because hypotonic fluid leaves.
4. **Mg^{2+} reabsorption** occurs in the loop of Henle.
5. The **Na^+-K^+-Cl^{2-} cotransporter** is located here and is affected by **loop diuretics** (eg, **furosemide**).
6. Flow through the loop of Henle is relatively slow, allowing the kidney to maintain a high medullary osmolarity.

BARTTER SYNDROME

- *Bartter syndrome is a kidney disease characterized by* ***Na^+, K^+, and Cl^- wasting.***
- ***Renin and aldosterone levels are increased,*** *but blood pressure remains low.*

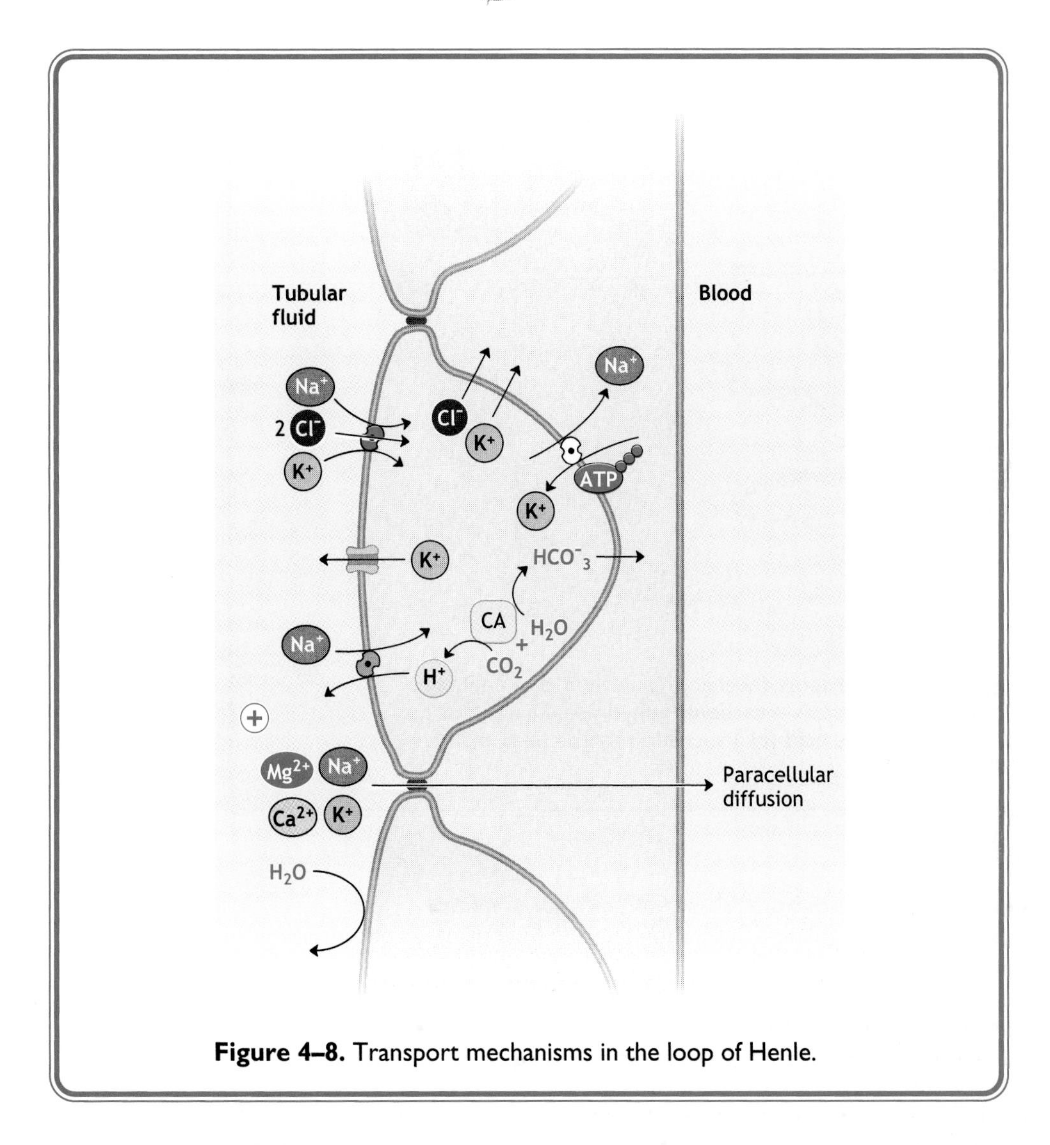

Figure 4–8. Transport mechanisms in the loop of Henle.

- *Symptoms include **polyuria** (excessive urination), **nocturia** (nighttime urination), **developmental delay,** and **dehydration.***
- *The primary defect is in Cl^- reabsorption in the ascending limb of the loop of Henle, which leads to decreased tonicity of the interstitium and an inability to concentrate urine.*
- *The syndrome results in **hypokalemic metabolic alkalosis** (described later in this chapter).*
- *Treatment is aimed at converting K^+ balance by oral potassium supplementation.*

C. Distal Nephron (Figure 4–9)

1. The **distal convoluted tubule** and the **collecting duct** reabsorb variable amounts of water depending on circulating levels of **antidiuretic hormone, (ADH,)** and **aldosterone.**
2. ADH stimulates increased water permeability in the distal convoluted tubule and the collecting duct, making the tubular fluid isosmotic with the ISF.

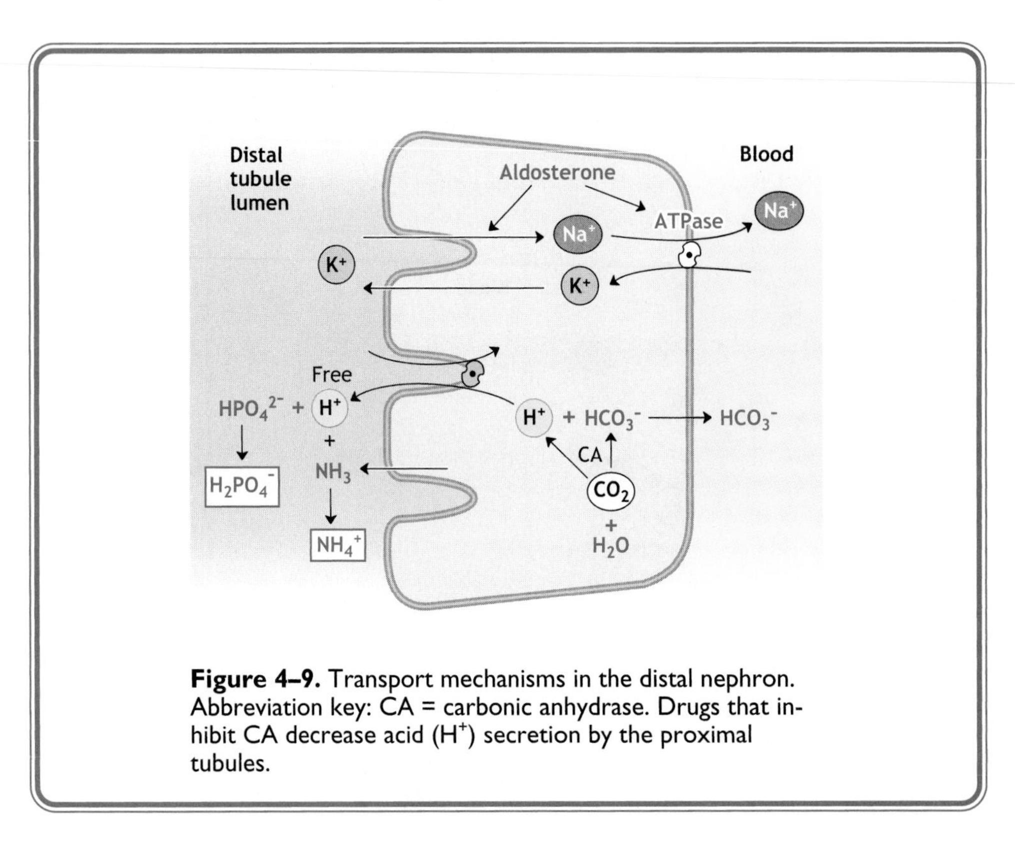

Figure 4–9. Transport mechanisms in the distal nephron. Abbreviation key: CA = carbonic anhydrase. Drugs that inhibit CA decrease acid (H^+) secretion by the proximal tubules.

3. **Aldosterone** increases Na^+ reabsorption and K^+ secretion.
4. These hormones regulate K^+ excretion and the final urinary concentrations of K^+, Na^+, and Cl^-.
5. Two main cell types are present in the distal nephron.
   a. **Principal cells** are involved with Na^+ and water transport.
   b. **Intercalated cells** secrete H^+ and reabsorb K^+.
6. These processes allow the kidney to secrete dilute or concentrated urine as necessary to maintain **homeostasis.**

7. The **early distal convoluted tubule** is the site of action of **thiazide diuretics,** which inhibit the Na^+-Cl^- cotransporter.

VI. Regulation of NaCl Transport

A. Because sodium salts are the predominant extracellular solutes, total body sodium determines extracellular fluid volume.
   1. Glomerulotubular balance stabilizes the fractional Na^+ reabsorption by the proximal tubule in the face of altered filleted Na^+ load.

B. The primary regulatory systems that respond to changes in body fluid volume are the **sympathetic nervous system,** the **renin-angiotensin-aldosterone system, atrial natruretic peptide** (**ANP**), and **ADH or AVP** (arginine vasopressin).
   1. The **sympathetic nervous system** has stretch receptors on blood vessels such as vena cava, cardiac atria, and pulmonary vessels.

a. A **decreased firing rate in the afferent nerves** from these volume receptors **increases sympathetic outflow from cardiovascular medullary centers.**
b. **Aldosterone, increased renal sympathetic tone,** and **arginine** vasopressin all increase Na^+ reabsorption.

2. **Aldosterone secretion** is controlled by the **renin-angiotensin system** (Figure 4–10).
   a. Granular cells in the wall of renal afferent arterioles, which are part of the juxtaglomerular apparatus, release **renin,** an enzyme that converts angiotensinogen from the liver to angiotensin I.
   b. **Renin** production is controlled by three mechanisms (Figure 4–11):
      (1) **Decreased sodium chloride delivery past macula densa cells** in the thick ascending limb of the loop of Henle increases renin release.
      (2) **Baroreceptors in the wall of the afferent arteriole** respond to decreased pressure, stretch, or shear stress by increasing renin release.
      (3) **Stimulation of β-adrenergic receptors** on the juxtaglomerular granular cells stimulates renin release.
   c. **Angiotensin I** is converted to angiotensin II by angiotensin-converting enzyme.
   d. The plasma level of renin is the rate-limiting step in the production of angiotensin II.
   e. **Angiotensin II** enhances salt retention by increasing **sodium reabsorption in the proximal tubule** as well as by its vasoconstrictor action, which reduces the GFR.
3. **Angiotensin II also stimulates aldosterone secretion** from the zona glomerulosa of the adrenal gland, **causing enhanced Na^+** reabsorption by the collecting duct.

CONN SYNDROME

- *Conn syndrome involves benign **adenoma** or hyperplasia **of the zona glomerulosa** of the adrenal cortex.*
- *It is caused by **primary hyperaldosteronism** due to unregulated production of aldosterone.*
- *Aldosterone increases distal Na^+ exchange for K^+ and H^+ ions, resulting in **hypernatremia** (excess blood sodium) and **hypokalemia** (low blood potassium).*
- ***Metabolic alkalosis** results from loss of H^+ ions, which increases reabsorption of HCO_3^-.*
- *Increased blood volume inhibits Na^+ reabsorption in the proximal tubule and ADH release.*
- *The excess reabsorption of Na^+ results in hypertension.*

4. **Natriuretic Factors such as ANP,** a peptide released by atrial distention that **inhibits Na^+ reabsorption** along the collecting duct, thereby causing **natriuresis.** ANF may also increase the GFR. ANF release is increased by volume expansion and decreased with volume depletion.
5. **Arginine Vasopressin or ADH** increases the water permeability of renal cells in the distal tubule and collecting duct, thus decreasing the volume and increasing the osmolarity of urine.
   a. **ADH is stimulated by intravascular volume depletion,** thereby promoting water retention.
   b. **ADH is synthesized in supraoptic and paraventricular nuclei of the hypothalamus** and is released from the posterior pituitary.

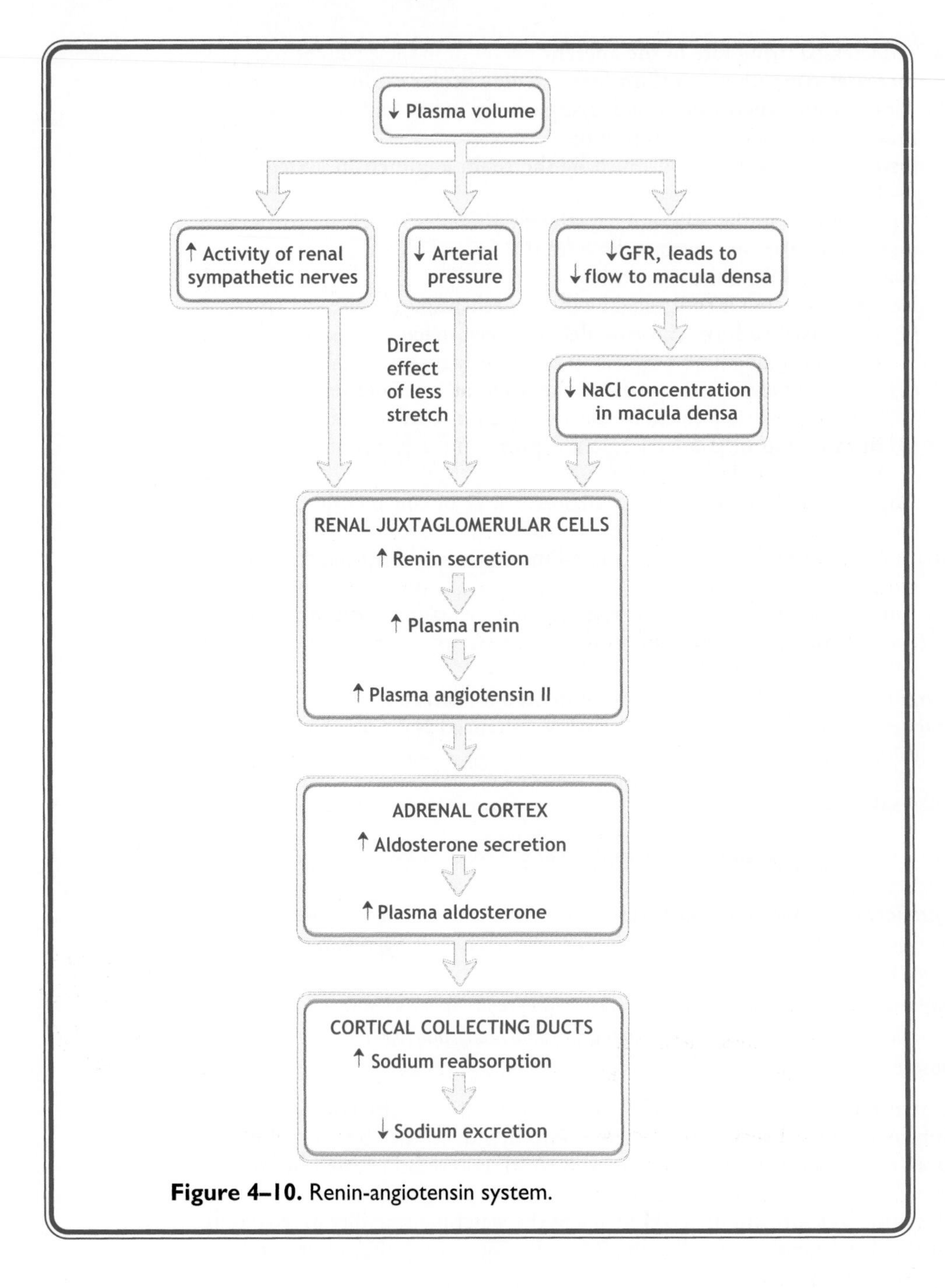

Figure 4–10. Renin-angiotensin system.

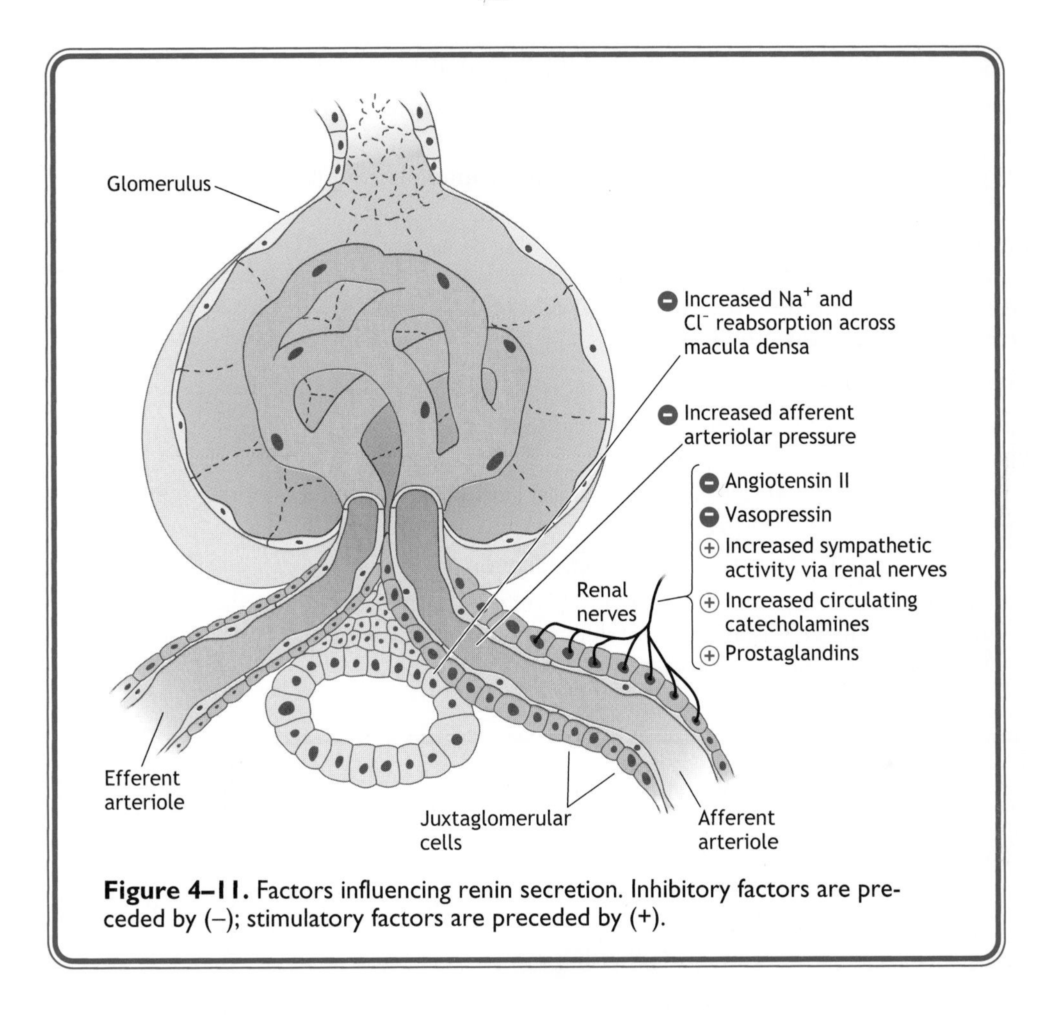

Figure 4–11. Factors influencing renin secretion. Inhibitory factors are preceded by (–); stimulatory factors are preceded by (+).

DIABETES INSIPIDUS

- ***ADH,*** *is* ***secreted into the blood*** *from the posterior pituitary gland.*
- *ADH increases the water permeability of the late distal tubule and collecting duct.*
- *Diabetes insipidus (DI) is a syndrome of ADH deficiency and is associated with* ***polydipsia*** *(excessive water intake) and polyuria.*
- ***Hypothalamic DI*** *results from a defect in the neural circuitry related to ADH synthesis and release.*
- ***Nephrogenic DI*** *is associated with a defect in the V_2 receptor gene or aquaporin 2 gene for ADH.*
- ***Polydipsic DI*** *is associated with compulsive water drinking.*

VII. Potassium Regulation

A. K^+ is filtered, reabsorbed, and secreted by the nephron.

B. Most of the filtered K^+ is reabsorbed in the proximal tubule.

C. Twenty percent is reabsorbed in the thick ascending limb of the loop of Henle, through its involvement in the **Na^+-K^+-Cl^{2-} cotransporter.**

D. K^+ balance is achieved when urinary excretion of K^+ equals dietary intake of K^+.

E. K^+ is passively secreted by the principal cells in the distal nephron via a K^+ channel and the K^+-Cl^{2-} cotransporter pathway.

F. Reabsorption of K^+ occurs in the distal nephron in the intercalated cells via the K^+-H^+ exchanger.

G. Increased K^+ excretion occurs in response to

1. **High intakes of K^+ or Na^+**
2. **Increased cell pH in the distal convoluted tubule**
3. **Increased plasma aldosterone levels**

VIII. Renal Handling of Glucose

A. The glucose-filtered load is directly proportional to the plasma glucose concentration.

B. The proximal tubule reabsorbs glucose via an apical electrogenic Na/Glucose cotransporter (SGLT) and a basolateral facilitated diffusion mechanism (GLUT).

C. The number of Na^+-glucose carriers is, however, limited.

D. At plasma glucose concentrations greater than 350 mg/dL, carriers are saturated; this is the Tm for glucose (Figure 4–12). Glucose excretion in the urine occurs only when the plasma concentration exceeds the Tm for glucose.

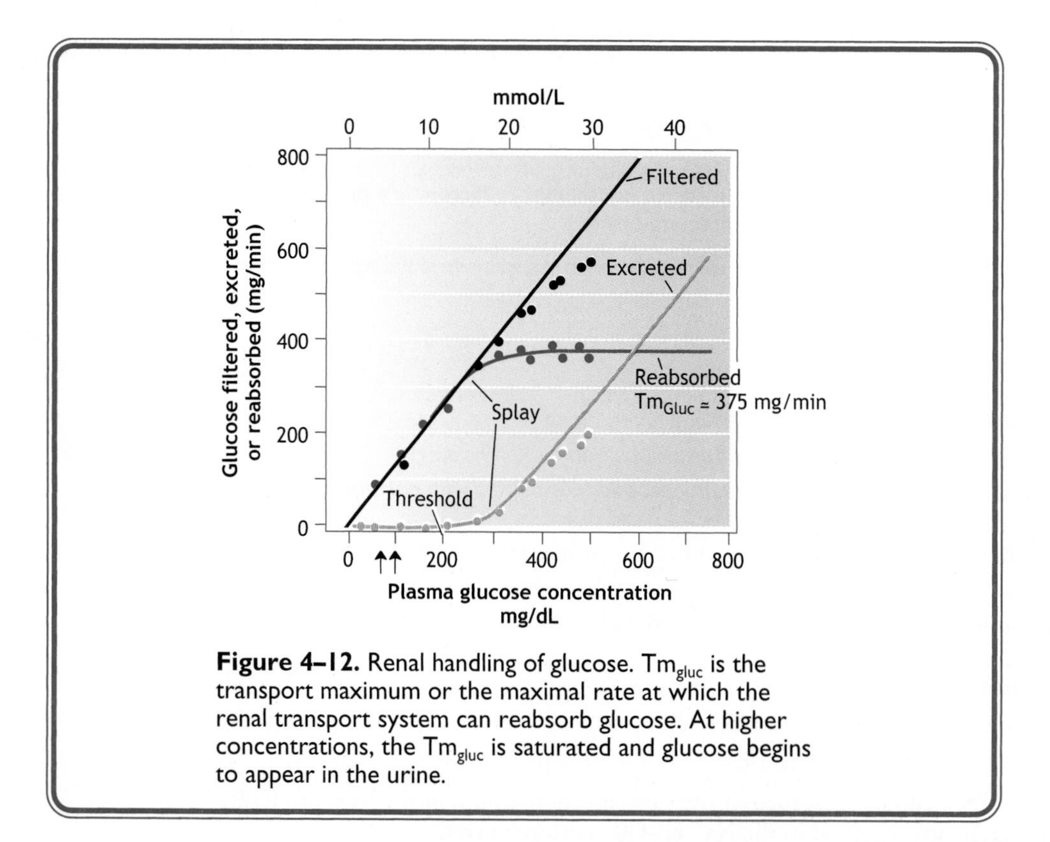

Figure 4–12. Renal handling of glucose. Tm_{gluc} is the transport maximum or the maximal rate at which the renal transport system can reabsorb glucose. At higher concentrations, the Tm_{gluc} is saturated and glucose begins to appear in the urine.

IX. Urea Regulation

A. **Urea,** an end-product of nitrogen metabolism, is an example of a **passively transported substance.**

B. The kidney filters, reabsorbs and secretes urea. About 50% of the filtered load **is reabsorbed in the proximal tubule.**

C. Although the distal tubule and collecting ducts are usually impermeable to urea, **ADH** increases the permeability of the medullary collecting ducts, thereby enhancing the osmolality of the medullary interstitium.

D. **Urea excretion coinsides with increased urinary flow.** Thus, in a water diuresis when ADH is low, the clearance of urea increases.

E. About 40% of the osmolality in the medulla is due to the presence of urea.

X. Phosphate Regulation

A. Most of the filtered phosphate is reabsorbed in the proximal convoluted tubule via an apical Na/phosphate cotransporter (NaPi).

B. **Parathyroid hormone** inhibits apical Na/phosphate uptake, causing phosphaturia (excess phosphate in urine).

C. **Phosphate is a buffer for H^+** in the urine and is excreted as H_2PO_4.

XI. Renal Calcium Regulation

A. **Ninety percent of the filtered calcium is passively reabsorbed** in the proximal tubule and thick ascending limb of the loop of Henle.

B. **Loop diuretics** (eg, **furosemide**) inhibit Ca^{2+} reabsorption because Ca^{2+} reabsorption is coupled with Na^+ reabsorption and blocked by loop diuretics.

C. **Thiazide diuretics** increase Ca^{2+} reabsorption in the distal tubule and collecting ducts and can be used to treat **hypercalcuria** (excess calcium in the urine).

D. **Parathyroid hormone** and **vitamin D increases** Ca^{2+} reabsorption in the distal tubule while high plasma Ca^{2+} inhibits calcium reabsorption.

XII. Magnesium Regulation

A. **Mg^{2+}** is primarily reabsorbed in the **thick ascending limb** of the loop of Henle.

B. **Mg^{2+}** and **Ca^{2+}** compete for reabsorption in the thick ascending limb.

C. **Hypercalcemia,** therefore, inhibits Mg^{2+} reabsorption, and **hypermagnesemia** inhibits Ca^{2+} reabsorption.

D. Mg^{2+} reabsorption increases with depletion of magnesium or calcium, extracellular fluid contraction and parathyroid hormone secretion.

E. Diuretics generally decrease Mg^{2+} reabsorption and thereby enhance Mg^{2+} excretion.

XIII. Concentrating and Diluting Mechanisms (Figures 4–13 and 4–14)

A. Generation of a corticomedullary osmotic gradient.

1. The purpose of the **countercurrent mechanism** is to **increase the osmolality of the interstitial fluid and concentrate urine.**
2. The countercurrent multiplication principle requires energy and differences in the membrane characteristics between the two limbs of the loop of Henle.
3. Active ion reabsorption by the thick ascending limb increases the interstitial osmotic gradient.

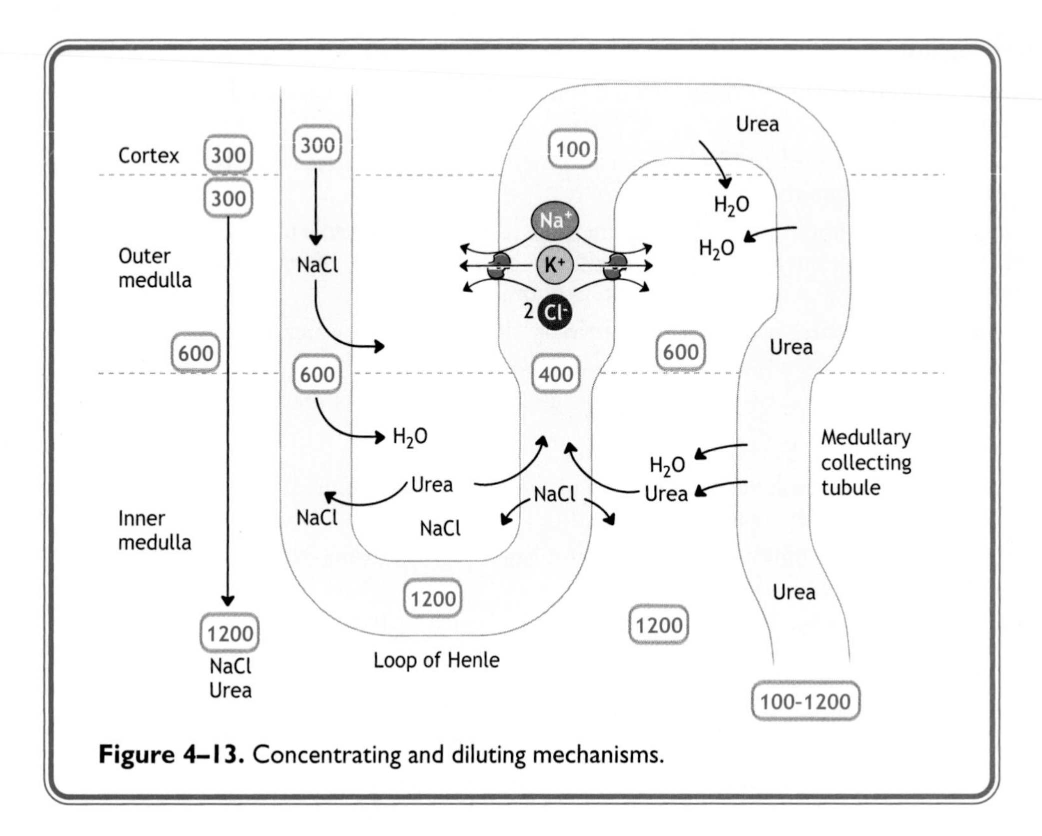

Figure 4–13. Concentrating and diluting mechanisms.

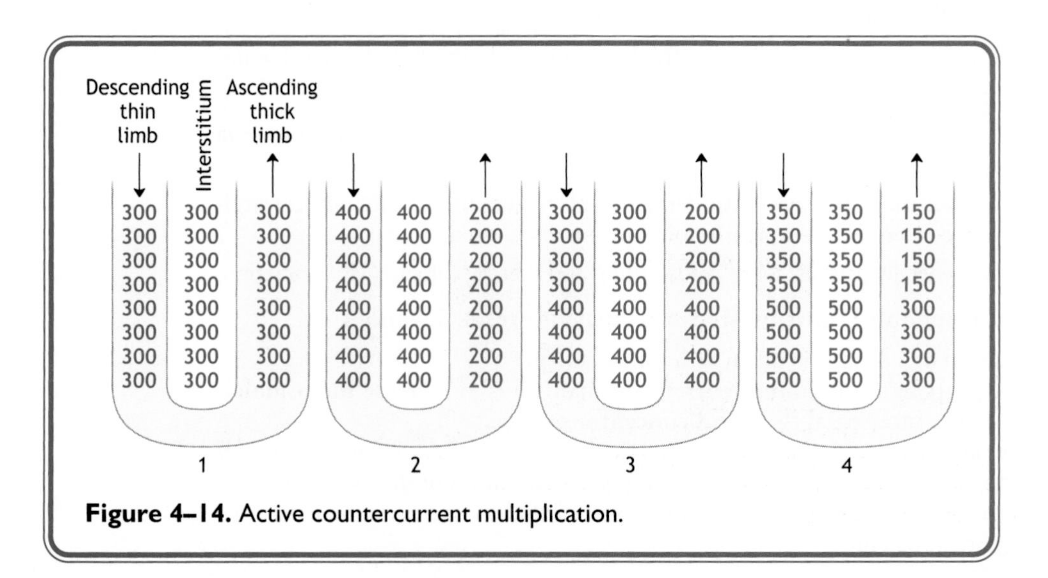

Figure 4–14. Active countercurrent multiplication.

4. Low water permeability of the ascending limb prevents dilution of the interstitial osmotic gradient.
5. High water permeability of the descending limb permits equilibration of contents with the interstitium.

B. The osmotic gradient (Figure 4–15) is maintained through
1. **Passive countercurrent exchange** in the vasa recta
2. **Low fluid flow rates** in the tubules and vasa recta
3. **Regulation of the permeability of collecting ducts to water and urea via ADH**

SYNDROME OF INAPPROPRIATE ADH SECRETION (SIADH)

- *SIADH is a common finding in patients with brain and lung lesions.*
- *The syndrome causes water retention and hyponatremia.*

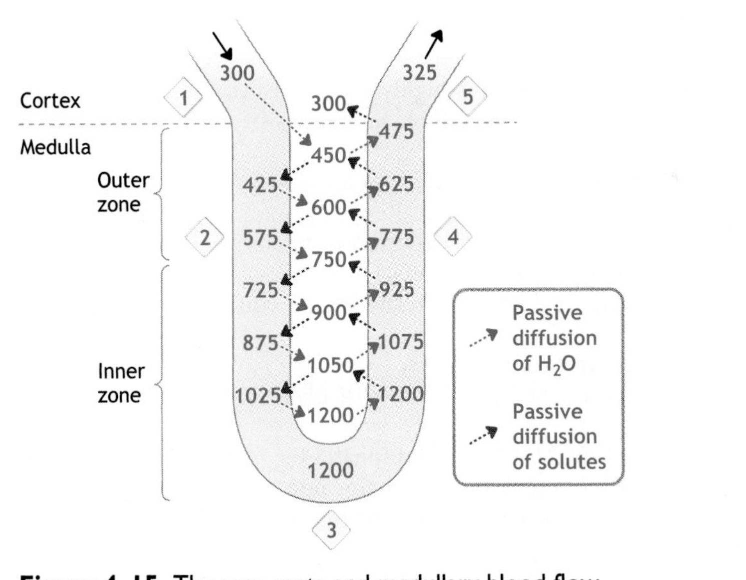

Figure 4–15. The vasa recta and medullary blood flow are essential for trapping ions in the renal medulla and papilla. The vasa recta are capillaries that allow free exchange between the blood and interstitial compartments and act as countercurrent exchanges. ***1.*** Water diffuses out of descending vessels and into ascending vessels. **2.** Solutes diffuse out of vessels ascending to cortex and into descending vessels passing to the medulla. **3.** A large osmotic gradient exists in the vascular loop area. ***4.*** Water diffuses into the ascending vessels, and solutes diffuse into the medullary interstitium. **5.** The osmolality of blood reentering the renal cortex is similar to that when it entered at ***1.***

- *The hematocrit remains unchanged because water shifts into the red blood cells, offsetting the gain of ECF volume.*
- *Treatment involves restricting free water intake to convert the inappropriate ADH secretion to normal levels via dehydration.*

XIV. Acid-Base Balance

A. Definitions

1. An **acid** is a **proton donor,** which is a molecule or ion that can contribute a hydrogen ion to a solution.
2. A **base** is a **proton acceptor,** which is a molecule or ion that will combine with a hydrogen ion to remove it from a solution.
3. H^+ is the **acid** and A^- is the **conjugate base** in HA, a **conjugate acid-base pair.**
4. The strength of an acid is defined with respect to the ease with which H^+ is released.
5. The strength of a base is defined with respect to how strongly it binds H^+.
6. For example, when two conjugate acid-base pairs such as HCl and H_2O interact, H_2O is the stronger conjugate base ($HCl + H_2O = H_3O + Cl$), binding H^+ more strongly than Cl^-.

B. Buffering Systems

1. H^+ concentration in body fluids is highly regulated because minor changes from the normal value can cause marked alterations in the rates of chemical reactions in the body.
2. **Buffers resist changes in pH** when H^+ ions are added to or removed from a solution.
3. A disturbance affecting the H^+ concentration is measured as a change in pH (ie, increased H^+ = decreased pH).
4. The capacity of a buffering system is related to its concentration and it pK (relative to pH).
5. The major **extracellular buffer** is $\mathbf{HCO_3^-}$.
6. **Phosphate** is a minor extracellular buffer that plays its most important role as a urinary buffer.
7. **Intracellular buffers** include organic phosphates (eg, ATP, ADP, and AMP) and proteins (of which **hemoglobin is a major one**).
8. The Henderson-Hasselbalch equation is used to calculate pH:

$$(pH = pK + \log \frac{[A^-]}{[HA]})$$

where
$pH = -\log_{10} [H^+]$
$pK = -\log_{10}$ equilibrium constant (pH units)
$[A^-]$ = base form of buffer (mM); H^+ acceptor
$[HA]$ = acid form of buffer (mM); H^+ donor

   a. If the molar ratio of A^- to HA and the pK of HA are known, the pH can be calculated.
   b. When the concentration of HA and A^- are equal, the pH of the solution equals the pK of the buffer.

Table 4–5. Primary acid-base disturbances.

Condition	Arterial Plasma			Cause
	pH	HCO_3^- (mEq/L)	P_{CO_2} (mmHg)	
Normal	7.40	24.1	40	
Metabolic acidosis	7.28	18.1	40	NH_4Cl ingestion
	6.96	5.0	23	Diabetic acidosis
Metabolic alkalosis	7.50	30.1	40	$NaHCO_3^-$ ingestion
	7.56	49.8	58	Prolonged vomiting
Respiratory acidosis	7.34	25.0	48	Breathing 7% CO_2
	7.34	33.5	64	Emphysema
Respiratory alkalosis	7.53	22.0	27	Voluntary hyperventilation
	7.48	18.7	26	3-week residence at 4000-meter altitude

C. Primary Acid-Base Disturbances (Table 4–5)

1. **Respiratory acidosis** is produced by **hypoventilation,** which increases CO_2 levels, resulting in a decrease in pH and a slight increase in HCO_3^-.
   a. Respiratory acidosis is diagnosed when **PCO_2 is greater than 40.**
   b. Other possible causes include **barbiturate overdose, pulmonary edema, ventilation-perfusion mismatch.**
2. **Respiratory alkalosis** is produced by **hyperventilation,** which decreases CO_2 levels, resulting in increased pH and a slight decrease in HCO_3^-.
   a. Respiratory alkalosis is diagnosed when **PCO_2 is less than 40.**
   b. Possible causes include **hyperventilation, high altitude, salicylate intoxication** (a few grams/day), **endotoxins,** and **anxiety.**
3. **Metabolic acidosis** is caused by a gain in fixed acid or a loss of HCO_3^- and decreased pH.
   a. Metabolic acidosis is diagnosed when **HCO_3^- is less than 24.**
   b. Possible causes include **diabetic ketoacidosis, renal failure, shock (lactic acidosis), and severe diarrhea** (HCO_3^- loss)
4. **Metabolic alkalosis** is caused by a loss in H^+ as fixed acid, which results in an increase in HCO_3^- and increased pH.
   a. Metabolic alkalosis is diagnosed when **HCO_3^- is greater than 24.**
   b. Possible causes include **severe vomiting** (when H^+ is lost) **diuretic abuse, and $NaHCO_3^-$ therapy.**

D. Serum Anion Gap (AG) and Metabolic Acidosis

1. Total cation changes in the plasma always equal the total anion changes.
2. The AG represents unmeasured ions (ie, protein, phosphate citrate, sulfate) in serum. Normal AG is 5–12 mEq/L.
3. **Metabolic acidosis is subdivided into increased AG and normal AG.** Increased AG is anything greater than 12 mEq/L.
4. If the fall in HCO_3^- is less than the rise in AG, **coexisting metabolic alkalosis** is suspected.
5. An AG of 12 means that 12 ions are unaccounted for (normally albumin, phosphate, and organic acids).
6. If AG is increased, then other ions (eg, phosphate, lactate, β-hydroxybutyrate) must be in the system to replace HCO_3^-.
7. **Increased AG** is most **useful in diagnosing the cause of metabolic acidosis** with the accumulation of organic anions, such as lactic acidosis, diabetic ketoacidosis, and the ingestion of sulfate.
8. **In type I renal tubular acidosis, H^+ cannot be secreted** in the distal tubule, **inhibiting HCO_3^- reabsorption** and promoting K^+-Na^+ exchange, which results in **hypokalemia.**
9. **In type II renal tubular acidosis, HCO_3^- reabsorption is defective** in the proximal tubule, resulting in an **increased negative charge** in **tubular urine,** depleting HCO_3^-, drawing out positive charged ions such as K^+, and causing **hypokalemia.**

E. Compensatory Mechanisms ($CO_2 \leftrightarrow H^+ + HCO_3^-$)

1. **Renal compensation for respiratory acidosis:** The primary defect is increased PCO_2 and reduced plasma pH. The kidney produces HCO_3^- and secretes it into the blood. For every HCO_3^- produced by the kidney, one H^+ will be excreted in the urine (producing **acidic urine**).
2. **Renal compensation for respiratory alkalosis:** The primary defect is reduced PCO_2 and elevated plasma pH. The kidney excretes HCO_3^- in the urine (producing **alkaline urine**). For every HCO_3^- excreted in the urine, one H^+ is returned to the blood. Plasma HCO_3^- decreases slowly as plasma H^+ increases.
3. **Respiratory compensation for metabolic acidosis:** Metabolic acidosis occurs when there is a decrease in the kidney's ability to excrete acid, most often manifested by a low GFR due to renal disease. **Hyperventilation** reduces CO_2 levels, shifting the reaction to the left and consuming H^+.
4. **Respiratory compensation for metabolic alkalosis:** Metabolic alkalosis occurs after prolonged vomiting with significant losses of HCl from the stomach. **Hypoventilation** increases CO_2 levels, shifting the reaction to the right and producing H^+.

XV. Diagnostic Hints for Acid-Base Disorders

A. Decreased pH indicates **acidosis.**

1. **Increased CO_2** indicates **respiratory acidosis.**
2. **Normal CO_2 with decreased HCO_3^-** indicates **metabolic acidosis.**
3. **Increased CO_2 with decreased HCO_3^-** indicates **combined respiratory and metabolic acidosis.**

B. Increased pH indicates **alkalosis.**
1. **Decreased CO_2** indicates **respiratory alkalosis.**
2. **Normal CO_2 with elevated HCO_3^-** indicates **metabolic alkalosis.**
3. **Decreased CO_2 with elevated HCO_3^-** indicates **combined respiratory and metabolic alkalosis.**

C. If **CO_2 and HCO_3^- change in opposite directions,** a **combined disturbance** is present.

METABOLIC ACIDOSIS

- ***Diabetes mellitus*** *is a major cause of metabolic acidosis.*
  *–**Type II diabetes mellitus** is the most common cause of ketoacidosis.*
  –Decreased insulin secretion leads to fat catabolism and ketoacidosis.
  *–Insulin deficiency is also associated with **hyperkalemia.***
  –Treatment of the primary disease (ie, insulin deficiency) corrects the disorder.
- ***Severe diarrhea*** *is another cause of metabolic acidosis.*
  –Small intestinal and colonic secretions are alkaline, containing a high concentration of HCO_3^-.
  –Significant HCO_3^- loss with prolonged diarrhea causes metabolic acidosis.
  *–Administration of NaHCO3 **is a useful treatment.***

METABOLIC ALKALOSIS

- ***Prolonged vomiting*** *is a primary cause of metabolic alkalosis and dehydration.*
- *Gastric secretions contain a high concentration of H^+ and Cl^-.*
- *Potassium depletion may also occur rapidly and presents the greatest danger. Gastrointestinal secretions contain K^+ in concentrations two to five times higher than in the ECF.*
- *The alkalosis is primarily due to loss of Cl^- from the plasma and not the loss of H^+ from the stomach.*
- ***Treatment*** *involves administration of **isotonic NaCl or KCl.***

RESPIRATORY ACIDOSIS

- *Acute respiratory acidosis can be caused by **asthma.***
- *Clinical features are referred to as CO_2 narcosis, which is characterized by cyanosis; fatigue; blurred vision; headache; and confusion that leads to delirium, convulsions, and coma.*
- ***Therapy is directed toward enhancing ventilation*** *through bronchodilators and steroids.*

RESPIRATORY ALKALOSIS

- ***Hypoxia at high altitude*** *or **severe anemia** may result in respiratory alkalosis.*
- *Clinical features include lightheadedness, altered consciousness, paresthesia (tingling, burning sensation) of the extremities, and tetany (hyperexcitability of muscles due to decreased extracellular ionized calcium).*
- ***Therapy*** *is directed toward **decreasing pulmonary gas exchange.** The paper-bag technique of increasing alveolar PCO_2 is effective.*

XVI. Selected Acid-Base Disorders

HYPONATREMIA WITH EDEMA (EG, NEPHROTIC SYNDROME)

- *Characterized by the inability of the glomerulus to filter protein, proteinuria and hypoalbuminemia.*
- *Edema involves an increase in hydrostatic pressure or a decrease in oncotic pressure.*
- *Alterations in Starling's forces cause a* ***transudate*** *to leak into the interstitial space, resulting in* ***pitting edema.***
- *Because most fluid is in the interstitial space, venous return is decreased, resulting in decreased cardiac output; decreased blood volume; and stimulation of renin-angiotensin, aldosterone, and ADH.*
- *Treatment involves restricting salt and water intake and treatment with diuretics such as furosemide to increase fluid loss.*

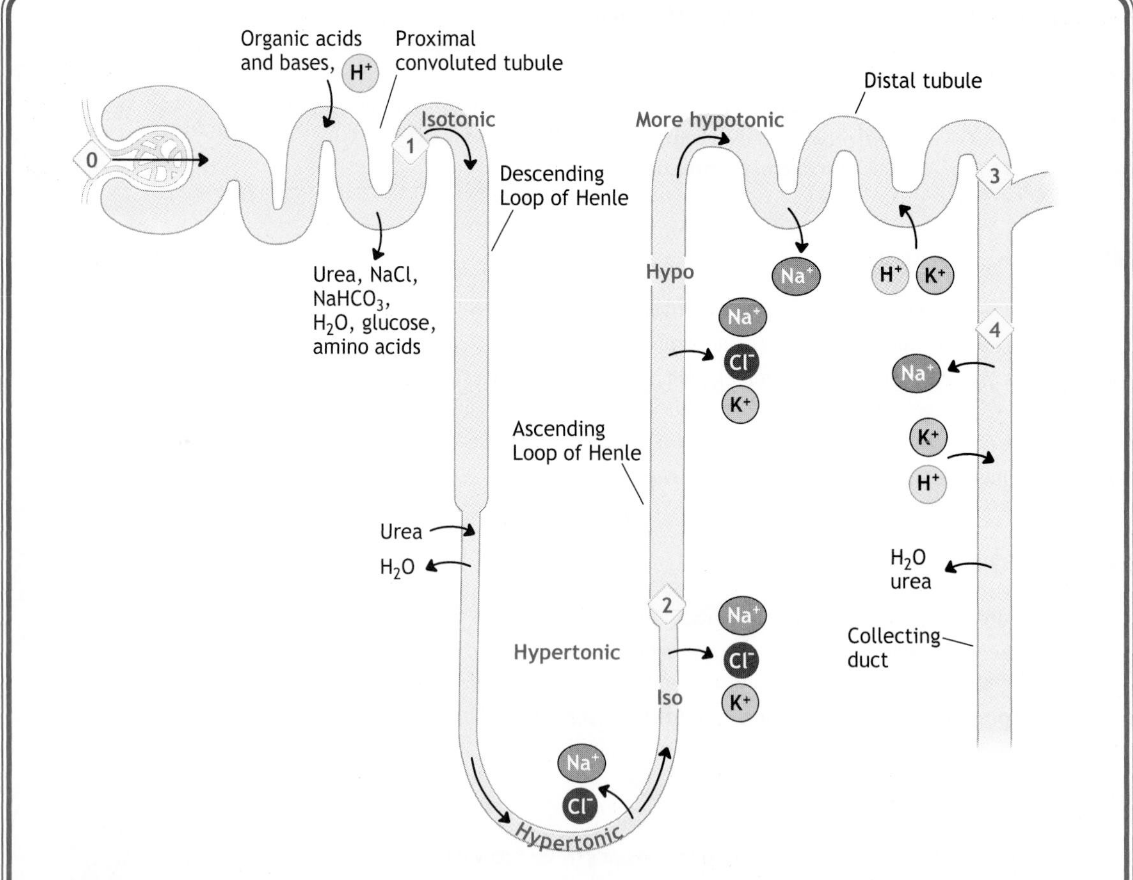

Figure 4–16. Osmotic diuretics (eg, mannitol) work in all parts of the nephron (***0***). Carbonic anhydrase inhibitors (eg, acetazolamide) block the acid secretion system in the proximal tubule (***1***). Loop diuretics (eg, furosemide) act on the thick ascending loop of Henle, which is impermeable to both water and urea (**2**). Thiazide diuretics (eg, hydrochlorothiazide) act on the distal convoluted tubule (**3**). Antagonists to aldosterone (eg, amiloride) and V_2 vasopressin receptor antagonists (eg, lithium) act on the collecting ducts (**4**).

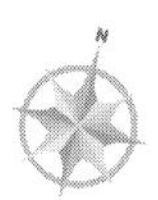

ADDISON DISEASE

- *Addison disease is caused by **adrenal insufficiency** resulting in deficient glucocorticoid and mineralocorticoid secretion.*
- *Lack of aldosterone results in **hyponatremia and hyperkalemia.***
- *Normal anion gap **metabolic acidosis** may develop from a primary loss of HCO_3^- due to hypoaldosteronism stemming from decreased mineralocorticoid activity.*
- ***Hypoglycemia** would be produced due to reduced glucocorticoid activity.*

DIURETIC EFFECTS (FIGURE 4–16)

CLINICAL CORRELATION

- ***Thiazide and loop diuretics** that block sodium reabsorption **cause a hypertonic loss of salt and water.***

- *Na^+ loss results in decreased circulating blood volume.*
- *Increased exchange of Na^+ for K^+ and H^+ results in **hypokalemia** and **metabolic alkalosis.***

POTASSIUM DISORDERS (FIGURE 4–17)

- ***Hypokalemia occurs in alkalosis,** when H^+ ions come out of cells and are then exchanged for K^+ inside the cells.*
- ***Hyperkalemia occurs in acidosis,** when excess H^+ ions enter cells and K^+ ions come out in exchange.*

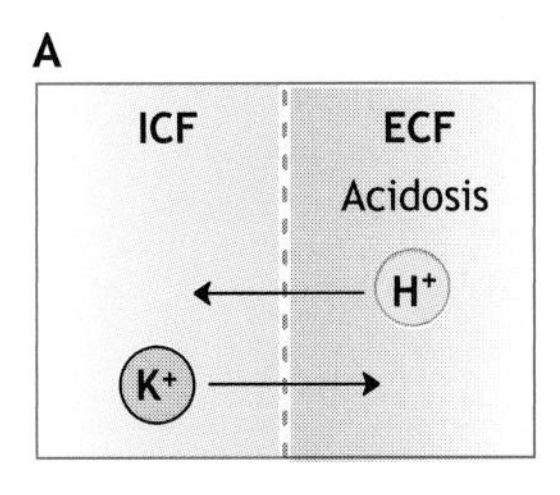

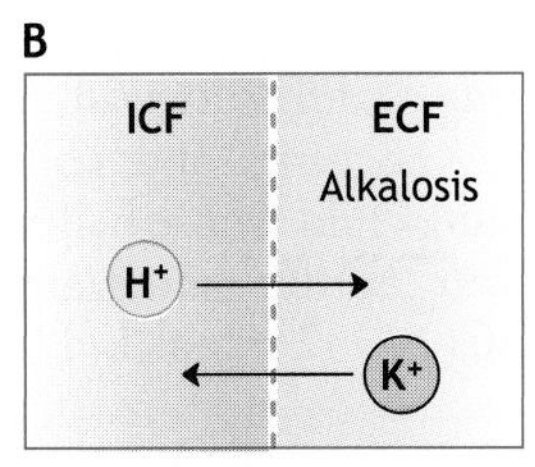

Figure 4–17. Renal handling of potassium. ***A.*** Hyperkalemia in alkalosis. ***B.*** Hypokalemia in alkalosis.

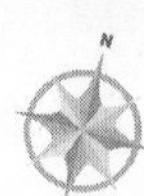

CLINICAL PROBLEMS

1. If a healthy 70-kg man loses 2 L of sweat while doing yard work and simultaneously drinks 2 L of pure water, which of the following body fluid changes would be expected?
   A. An increase in extracellular osmolarity
   B. An increase in extracellular fluid volume
   C. An increase in intracellular osmolarity
   D. An increase in intracellular fluid volume
   E. An increase in plasma Na+ concentration

A patient has 40 L of intracellular fluid (ICF) and 20 L of extracellular fluid (ECF). One and a half liters of a 0.15-M NaCl solution is infused intravenously, and after 1 hour there is complete equilibration with negligible excretion.

2. Which of the following ICF and ECF volume changes would be observed?
   A. ICF = +2.0 L; ECF = −0.5 L
   B. ICF = +1.5 L; ECF = no change
   C. ICF = +1.0 L; ECF = +0.5 L
   D. ICF = no change; ECF = +1.5 L
   E. ICF = −1.0 L; ECF = +2.5 L

A 60-kg man exhibits the following volume of distribution of tritiated water (THO): THO, 35 L; RISA, 3 L; and inulin, 8 L, after suitable time for mixing.

3. What is the subject's ISF volume?
   A. 2 L
   B. 4 L
   C. 5 L
   D. 8 L
   E. 10 L

A 6-year-old girl is brought to your office complaining of difficulty walking and weakness. In addition, she has been experiencing polydipsia, nocturia, and polyuria for several months. Physical examination reveals a healthy-looking child whose height and weight are between the 5th and 10th percentiles. The following serum values are obtained: Na^+, 136 mEq/L; K^+, 2.8 mEq/L; Cl^-, 90 mEq/L; and HCO_3^-, 32 mmol/L. Plasma renin levels are elevated. Urine screening for diuretics is negative.

4. Which of the following conditions is most consistent with the above data?
   A. Conn syndrome (primary hyperaldosteronism)
   B. Chronic licorice ingestion
   C. Bartter syndrome

D. Wilms tumor

E. Secondary hyperaldosteronism

A 19-year-old male visits your office complaining of polyuria and polydipsia. The following serum levels are obtained: Na^+, 144 mEq/L; K^+, 4.0 mEq/L; Cl^-, 107 mEq/L; and HCO_3^-, 25 mEq/L. Urine osmolality is 195 mOsm/kg water. Following 12 h of fluid deprivation, body weight has fallen 5%. Urine electrolytes are as follows: Na^+, 24 mEq/L; and K^+, 35 mEq/L. One hour later, the patient was infused with 5 IU of pitressin (ADH) that results in no change in his urine osmolality and electrolytes.

5. Which of the following is the likely diagnosis?

   A. Nephrogenic diabetes insipidus

   B. Osmotic diuresis

   C. Salt-losing nephropathy

   D. Psychogenic polydipsia

   E. Central diabetes insipidus

A patient is given a drug that causes an increased volume of urine with an osmolality of 100 mOsm/L.

6. This drug

   A. Inhibits renin secretion

   B. Decreases the active transport of Cl^- by the ascending limb of the loop of Henle

   C. Increases water permeability of the collecting duct

   D. Inhibits ADH secretion

   E. Increases the GFR

A patient with cirrhosis and ascites has been treated aggressively with a potent diuretic (eg, furosemide). After a few days, he experiences symptoms of weakness, muscle cramps, postural dizziness, and mental confusion. After hospitalization, the following laboratory values are obtained: plasma Na^+, 137 mEq/L; plasma K^+, 2.5 mEq/L; arterial pH, 7.58; and PCO_2, 50 mmHg.

7. Which of the following is a likely diagnosis?

   A. Respiratory alkalosis without renal compensation

   B. Chronic respiratory alkalosis with considerable renal compensation

   C. Metabolic alkalosis without respiratory compensation

   D. Metabolic alkalosis with some respiratory compensation

   E. Diabetes insipidus

A middle-aged woman has had asthma since childhood and has been a heavy smoker since her early teens. During the past few years, she has experienced progressive dyspnea (breathing difficulty) and somnolence (sleepiness). Physical examination reveals a cachectic (general ill health and malnutrition) female with shortness of breath, prolonged expirations, and frequent coughing. Laboratory data are as follows: arterial pH, 7.35; arterial HCO_3^-, 32 mEq/L; arterial PCO_2, 60 mmHg; and arterial PO_2, 60 mmHg.

8. Which of the following is a likely diagnosis?
   A. Acute metabolic acidosis with renal compensation
   B. Acute respiratory acidosis without renal compensation
   C. Chronic metabolic acidosis with considerable renal compensation
   D. Chronic respiratory acidosis with considerable renal compensation
   E. Respiratory acidosis with metabolic acidosis

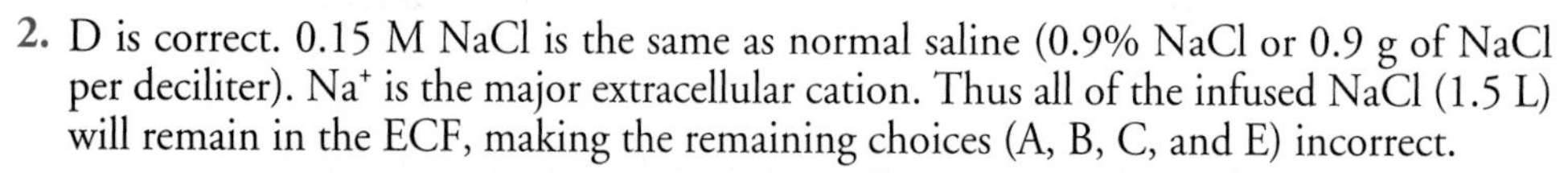

ANSWERS

1. D is correct. Two liters of sweat containing NaCl has been lost. ECF has been lost. Because Na_+ is primarily an extracellular cation, ingested pure water is hypotonic to the ICF and will be drawn into the more hypertonic intracellular medium, increasing ICF volume. An increase in extracellular osmolarity (choice A) is incorrect because loss of NaCl in sweat and its replacement with pure water will not replenish the decreased extracellular osmolality. Increased ECF volume (choice B) is incorrect because drinking pure water after sweating will cause most of the water to move into the ICF volume. Increased intracellular osmolarity (choice C) is incorrect because pure water moves into the intracellular space to make it more hypotonic and increase the ICF volume. Increased plasma Na^+ concentration (choice E) is not correct because Na^+ is lost in sweat and is not replaced by drinking water.

2. D is correct. 0.15 M NaCl is the same as normal saline (0.9% NaCl or 0.9 g of NaCl per deciliter). Na^+ is the major extracellular cation. Thus all of the infused NaCl (1.5 L) will remain in the ECF, making the remaining choices (A, B, C, and E) incorrect.

3. C is correct. RISA is used to measure plasma volume, which equals 3 L. Inulin is used to measure ECF volume, which equals 8 L. To measure interstitial fluid volume (ISF) one subtracts the plasma volume (PV) from the ECF. ISF + PV = ECF; therefore, ECF − PV = ISF, or 8 − 3 = 5 L.

4. C is correct. Bartter syndrome is characterized by Na^+, Cl^-, and K^+ wasting along with elevated renin and aldosterone levels. Laboratory values in this case indicate reduced values in the ions. In addition, renin levels are elevated, which leads to increased aldosterone levels. Patients with Bartter syndrome experience chronic volume depletion due to a defect in Na^+-Cl^- reabsorption in the thick ascending limb of the loop of Henle. Because urine screening for diuretic abuse is negative, the K^+ wasting points to Bartter syndrome. Conn syndrome (primary aldosteronism) (choice A); ingestion of licorice (choice B), which contains glycyrrhizinic acid, a substance that mimics the action of aldosterone; and secondary hyperaldosteronism (choice E) would all result in elevated serum Na^+. Wilms tumor (choice D) is not correct because patients with these tumors are hypertensive and have elevated renin levels.

5. A is correct. Nephrogenic diabetes insipidus is characterized by the inability of the kidney to respond to circulating vasopressin and retain water. The nephrogenic origin is indicated by a lack of response in urine concentration to exogenous ADH. Osmotic diuresis (choice B) is incorrect because the patient's urine osmolality (195 mOsm/kg of

water) did not change after ADH infusion. A hypotonic urine is expected in osmotic diuresis. Salt-losing nephropathy (choice C) and psychogenic polydipsia (choice D) are incorrect because the patient does not have hyponatremia and exhibited no response to ADH infusion. Central diabetes insipidus (choice E) is incorrect because there was no increase in urine osmolality after ADH administration, indicating that the polyuria and polydipsia were not caused by a lack of ADH.

6. D is correct. The drug increases the volume of hypotonic urine, thus inhibiting ADH secretion. Inhibition of renin secretion (choice A) is incorrect because loss of renin would lead to decreased Na^+ and water retention, which would cause increased urine osmolality. Decreased Cl^- transport (choice B) would lead to a more hypertonic urine, not a hypotonic urine. Increased water permeability of the collecting duct (choice C) decreases urine volume and, therefore, is incorrect. Increases in the GFR (choice E) would not necessarily increase or decrease the urine volume and is not relevant to this question.

7. D is correct. Metabolic alkalosis with partial respiratory compensation is identified through the increased arterial pH along with increased PCO_2. Alkalemia is associated with hypokalemia, as seen in this case. The metabolic alkalosis may result from the reduction of ECF volume due to diuretic administration. Respiratory alkalosis without renal compensation (choice A) is incorrect because hypocapnia is observed in respiratory alkalosis but was not observed in this case. Chronic respiratory alkalosis with considerable renal compensation (choice B) is incorrect because normal potassium levels are observed in chronic respiratory alkalosis. Metabolic alkalosis without respiratory compensation (choice C) is incorrect because PCO_2 levels are elevated in this case, indicating some respiratory compensation. Diabetes insipidus (choice E) is incorrect because it is characterized by dilute urine with hypernatremia, not by the normal sodium levels in this case.

8. D is correct. Chronic respiratory acidosis with considerable renal compensation is indicated by the arterial pH being only slightly acidic despite elevated CO_2 levels. The patient's history indicates severe chronic obstructive pulmonary disease (COPD) due to chronic asthma. The laboratory data indicate hypercapnia that is associated with HCO_3^- generation by the kidney. Due to obstructive disease, the patient has increased CO_2 production, and the patient's lung problem of poor alveolar ventilation enhances CO_2 retention. The increased CO_2 retention is associated with the observed hypoxemia. Acute metabolic acidosis with renal compensation (choice A) and acute respiratory acidosis without renal compensation (choice B) can be eliminated based on the patient's long history of COPD, indicating a chronic, not an acute, problem. Chronic metabolic acidosis with considerable renal compensation (choice C) is incorrect because in metabolic acidosis, HCO_3^- levels and PCO_2 levels are increased, not decreased. Respiratory acidosis with metabolic acidosis (choice E) is incorrect because although evidence of respiratory acidosis is present (increased CO_2 with secondary increases in HCO_3^- levels), there is no evidence for metabolic acidosis (decreased HCO_3^- with secondary decreases in PCO_2 levels).

CHAPTER 5
GASTROINTESTINAL PHYSIOLOGY

I. Regulation: Muscle, Nerves, and Hormones of the Gut

A. Muscles of the gut deal with movement and mechanical processing of luminal contents—moving, mixing, and storing ingested food.

B. **Voluntary muscle** is located at the upper (mouth, pharynx, and first third of the esophagus) and lower (external anal sphincter) gastrointestinal (GI) tract.

C. **Smooth muscle** structures have a nervous system of their own that can function without any extrinsic innervation (Figure 5–1).

D. This **enteric nervous system** (**ENS**) coordinates all activities and consists of the myenteric plexus between the longitudinal and circular muscle layers and the submucosal plexus between the circular muscle and muscularis mucosa.

1. The ENS is a "minibrain" with sensory neurons, interneurons, and motor neurons.
2. Receptors in the wall of the gut may be **chemoreceptors** that respond to chemicals such as hydrogen ions or **mechanoreceptors** that respond to stretch or tension.
3. Efferent fibers connect with muscles to cause contraction, with endocrine cells to release peptides, and with secretory cells to release secretions.
4. The brain-gut axis controls GI function via the autonomic nervous system, GI hormones and the immune system.
   a. The mucosa of the gastric antrum and the small intestine contains primarily endocrine cells.
   b. There are four major **regulatory peptides** in the gut:
      (1) **Gastrin** is released from the gastric antrum G cells by stomach distention, vagal innervation, and protein digestive products. It stimulates gastric secretion, motility, and mucosal growth.
      (2) **Cholecystokinin** (**CCK**) is released by duodenal I cells stimulated by fat and amino acids. CCK stimulates pancreatic enzyme secretion and contraction of the gallbladder primarily.
      (3) **Secretin** is released by acid from the S cells of the duodenum. It stimulates HCO_3^- secretion from the pancreas and liver, and inhibits gastric motility and secretion.
      (4) **Gastric inhibitory peptide, or glucose insulinotropic peptide** (**GIP**), is released by dietary fat, carbohydrate, and amino acids (from duodenal cells). It stimulates insulin release and inhibits gastric motility and secretion.

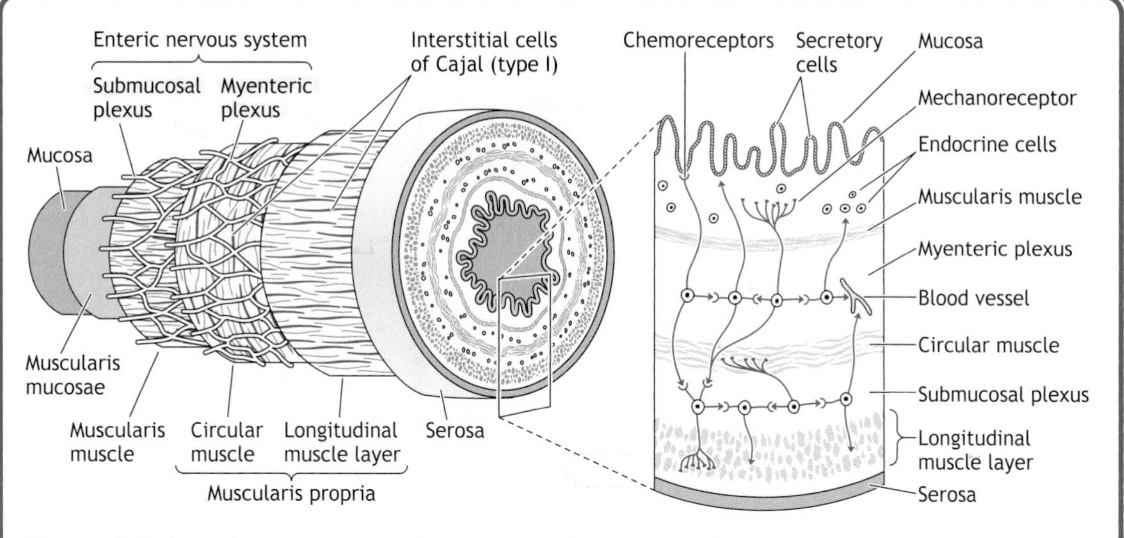

Figure 5–1. Smooth muscle lies between the two ends of the gastrointestinal tract and is arranged in three layers—outer longitudinal, inner circular, and muscularis mucosa—with all layers functioning as a unit.

E. Although the whole system can function without extrinsic innervation, extrinsic parasympathetic fibers are generally responsible for cholinergic and excitatory effects, and sympathetic fibers are associated with adrenergic and inhibitory effects.

F. **Contraction and relaxation of GI smooth muscle** is related to the calcium content of smooth muscle cells; increased cytosolic calcium causes contraction and vice versa.

II. Salivary Secretion

A. Anatomic Considerations

1. Between 1 and 1.5 L of saliva per day is produced by continuous secretion of the three salivary glands.
2. **Salivary secretion** is a composite of the three salivary gland secretions:
   a. The **parotid gland** generates 25% of the total secretion and is composed of serous cells that produce watery secretions.
   b. The **submandibular gland** accounts for 70% of the total secretion and produces mucous (protein) and serous secretions.
   c. The **sublingual gland** contributes 5% of the total secretion and produces mainly mucous (protein) secretions.
3. Anything in the mouth increases secretions via afferents stimulating the salivation center.

B. Inorganic Constituents of Secretions

1. The inorganic and organic **constituents of salivary secretions form a hypotonic secretion** because salivary ducts are impermeable to water.
2. The **basic electrolytes** in saliva include Na^+, Cl^-, HCO_3^-, and K^+ (Figure 5–2).

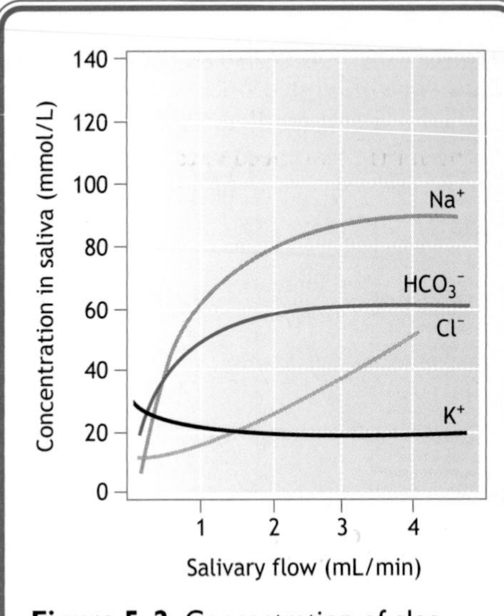

Figure 5–2. Concentration of electrolytes in saliva. *(Adapted from Thaysen JH, Thorn NA, Schwartz IL. Excretion of sodium, potassium, chloride, and carbon dioxide in human parotid saliva. Am J Physiol 1954;178:155.)*

a. Salivary ductal cells produce a hypotonic fluid low in NaCl and high in $KHCO_3$.
b. At high rates of saliva secretion, there is not enough time for normal absorption to occur. Thus, greater amounts of Na^+, Cl^-, and HCO_3^- appear in the saliva.
c. Aldosterone, a mineralocorticoid, increases Na^+ reabsorption while parasympathetic stimulation decreases Na^+ absorption by ductal cells.

C. Organic Constituents of Secretions

1. **Salivary acinar cells secrete different proteins.**
2. **Ptyalin,** a salivary **α-amylase,** attacks the α1–4 glucosidic linkages of starch, resulting in maltose, maltotriose, and α-limit dextrins. Ptyalin continues to work in the stomach as long as the bolus of food remains intact, even if the optimum pH for amylase functioning (ie, 6.9) is not maintained.
3. **Lingual lipase** initiates fat digestion.
4. **Kallikrein** is an enzyme that splits off vasodilating protein (such as bradykinins) from the plasma. If saliva is injected into an animal, the vasodilatory properties of the saliva cause a drop in the recipient's blood pressure.
5. **Sex steroids** are also secreted in saliva.
   a. The salivary glands excrete **testosterone;** therefore, salivary testosterone levels can indicate male endocrine status.
   b. **Estrogen** and **progesterone** are also excreted in saliva.
6. **Mucins** are glycoproteins that lubricate and protect oral mucosa.

D. Functions of Salivary Secretion

1. **Digestion:** Salivary amylase initiates the breakdown of starch. Amylase functions optimally at a pH of 6.9 and is inhibited once it reaches the low pH (~3.9) of the stomach. Lingual lipase begins fat digestion.
2. **Lubrication:** Mucins provide the lubrication needed to facilitate speech and swallowing.
3. **Water balance:** When body water tables are low, the mouth becomes dry, stimulating thirst.
4. **Protection:** Saliva performs a cleansing function aided by immunoglobulin A, lysozymes, thiocyanate, lactoferrin, and HCO_3^-. HCO_3^- helps neutralize acid refluxed from the stomach and inhibits dental cavity formation by neutralizing acid produced by bacteria acting on food.
5. **Endocrine:** Endocrine steroids and peptides appear in saliva in amounts that reflect plasma levels. Thus, sex steroids found in the saliva can aid in the diagnosis of hypogonadism. **Vasoactive intestinal peptide** (**VIP**) and **epidermal growth factor** (**EGF**) are also present in saliva. EGF is associated with tooth eruption, maturation of the cellular lining of the gut, and cytoprotection of the esophagus.
6. **Excretory:** Substances are excreted out of the saliva. Certain symptoms may indicate the presence of poisons or viruses in saliva (eg, blue gums are diagnostic for lead poisoning).

E. Regulation of Secretion

1. The **nervous system controls secretion.**
2. The salivary center is in the 4th ventricle and receives input from the limbic system.
3. Sympathetic stimulation results in vasoconstriction and increased secretion of thick, viscous saliva.
4. Parasympathetic stimulation by cranial nerves VII, IX, and XII results in a copious, watery secretion.
5. Excessive salivation occurs prior to vomiting. The medullary vomiting center and salivation center are located close together in the medulla.

HYPERSALIVATION AND HYPOSALIVATION

CLINICAL CORRELATION

- ***Sjögren Syndrome*** *is a chronic autoimmune disease that affects salivary secretion primarily in women.*
- *Patients generate antibodies that react with their salivary glands to cause loss of the Cl^--HCO_3 exchanger.*
- *Symptoms include dry mouth, loss of taste, and difficulty with speech and swallowing.*
- *No specific treatment is available.*
- *Diminished salivation in* ***gastroesophageal reflux disease*** *(**GERD**) decreases the neutralizing capacity of saliva, resulting in esophagitis. Smoking contributes to hyposalivation.*
- *Excessive salivation is produced by anticholinesterase drugs (neostigmine) and insecticides (parathion).*

III. Swallowing

A. Swallowing is coordinated by the **medullary swallowing center,** which is stimulated by sensory input from the mouth via cranial nerves V, IX, and X.

B. Once initiated by the movement of food to the rear of the mouth, the sequence proceeds to completion through efferent messages to muscles of the mouth, pharynx, and esophagus.

1. The **oropharyngeal phase** is characterized by movement of food to the rear of mouth, elongation of the soft palate to close off the nasopharynx, inhibition of respiration, tipping over of the epiglottis to block the airway, upward movement of the hyoid bone and larynx, and relaxation of the upper esophageal sphincter.
2. The **esophageal phase** is characterized by a primary peristaltic wave that pushes the bolus toward the stomach, and relaxation of the **lower esophageal sphincter** (**LES**) allows food to enter the stomach. A secondary peristaltic wave clears residual material left behind.

C. The **LES** is a **barrier to the reflux of the stomach contents** into the esophagus and thus in the resting state maintains a pressure higher than in the stomach.

1. Foods that decrease LES pressure include chocolate, peppermint, and alcohol; high-protein meals increase LES pressure.
2. Important hormones that decrease LES pressure include progesterone, a female sex steroid present at higher levels during pregnancy and the luteal phase of the menstrual cycle, and CCK, a GI peptide released from the small intestine in response to fat and protein meals.
3. The **contraction and relaxation** of the LES is **mediated by neurotransmitters:** acetylcholine, which causes LES contraction, and VIP and nitric oxide (NO), which cause LES relaxation.
4. Thus, parasympathetic innervation of the LES is both excitatory (through acetylcholine release) and inhibitory (through VIP and NO release).

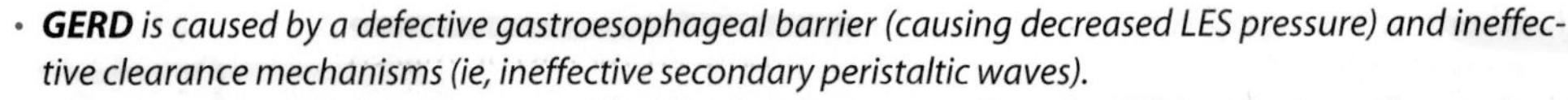

ESOPHAGEAL MOTOR DYSFUNCTION

- ***GERD** is caused by a defective gastroesophageal barrier (causing decreased LES pressure) and ineffective clearance mechanisms (ie, ineffective secondary peristaltic waves).*
  *–Chronic acid reflux damages mucosa leading to inflammation (**esophagitis**) and eventually to columnar epithelium replacement of squamous epithelium (**Barrett esophagus**), a precancerous condition.*
  –Lifestyle modifications that can prevent damage include elevation of the head of the bed, loss of excess weight, and avoidance of foods that lower LES pressure.
  –Medications include antacids to neutralize acid, histamine (H_2) receptor blockers to decrease acid secretion, proton pump inhibitors to stop acid secretion, and parasympathomimetic drugs that increase LES pressure (eg, methacholine).
- ***Achalasia** is a disease in which the LES fails to relax and esophageal peristalsis is absent. It is characterized by pain upon eating or drinking.*
  –Although the exact cause remains unknown, symptoms are thought to be due to an absence of inhibitory neurons in the esophageal intrinsic plexus.
  *–The most effective treatment for this condition involves **pneumatic dilation,** in which high air pressure stretches the constricted LES muscles to induce relaxation.*
  –Pharmacologic intervention, consisting of anticholinergics, nitrates, and calcium channel blockers can be used to relax the LES.
  *–**Esophagomyotomy, or Heller procedure,** a surgical procedure in which the longitudinal muscle is cut to induce relaxation, is also used.*

IV. Gastric Motor Function

A. Fed Motor Pattern

1. After a meal, peristaltic waves move toward the antrum to the pyloric sphincter, slowly propelling the mixture of food and gastric acid into the duodenum.

a. **Peristalsis** is controlled by a wave of partial depolarization known as the **basic electrical rhythm** (**BER**) or slow wave.

b. The BER begins in a group of pacemaker cells in the greater curvature and sweeps over the outer longitudinal muscle toward the pylorus.

(1) The BER may or may not be accompanied by contraction of underlying circular muscle.

(2) For example, when vagal fibers are activated by distention of the stomach, circular muscle fibers are depolarized enough to bring them to threshold so that they have action potentials and contraction occurs.

(3) Contractions of circular muscle occur in step with the BER-induced depolarization wave moving over the antrum.

(4) Gastric waves occur only when BER depolarizations reach the threshold for action potential discharges.

(5) A BER reaching threshold is determined by a combination of stretch, neural (vagal), and humoral (gastrin) stimuli.

2. The three major gastric motor activities of the fed stomach include **receptive relaxation, mixing,** and **emptying.**

a. With each swallow, the proximal stomach stretches to receive food from the esophagus, which involves only a small rise in intragastric pressure (**receptive relaxation**).

b. Receptive relaxation of the proximal stomach is a vagally mediated reflex.

c. The distal stomach grinds and mixes food to reduce bolus size so that it can be moved to the small intestine through the pyloric sphincter.

d. Muscle contractions of the antrum control the amount of food that leaves the stomach so as not to overload the digestive ability of the small intestine.

e. The amount of **chyme** (semi-fluid material produced by gastric digestion of food) emptied depends on the strength of the peristaltic wave and the pressure gradient between the antrum and duodenum.

f. The **pylorus** limits the size of particles emptied and acts to prevent reflux of duodenal contents into the stomach.

g. The volume and composition (ie, **osmolality, pH,** and **caloric content**) of gastric contents influence gastric emptying.

B. Fasting Motor Pattern: Migrating Motor Complex (MMC)

1. The MMC is the **pattern of a fasting or interdigestive state** that is divided into three phases (Figure 5–3).

2. The MMC moves stomach contents through the intestine to the ileocecal valve during overnight fasting.

3. The MMC performs a housekeeping function by sweeping gastric acid to the ileum to prevent bacterial overgrowth in the gut.

4. The GI regulatory peptide, motilin, is associated with initiation of MMCs in the stomach.

5. Feeding interrupts MMC activity by unknown causes.

C. Control of Gastric Emptying

1. Volume: Emptying of isotonic, noncaloric fluids is proportional to the volume or distention of the stomach.

2. Osmolality: Hypertonic and hypotonic fluid empty more slowly than isotonic fluids, probably because of neural and hormonal factors.

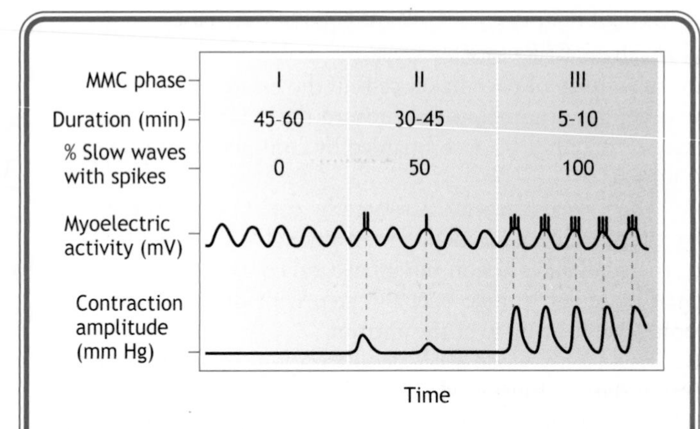

Figure 5–3. The fasting motor pattern has three phases, illustrated by the migrating motor complex (MMC). Phase I is the quiescent period, lasting 45–60 minutes. In phase II, which lasts 30–45 minutes, 50% of slow waves are associated with contractions. In phase III, 100% of slow waves are associated with strong contractions. Although this phase lasts only 5–10 minutes, gastric material is moved large distances.

3. **pH:** The lower the pH, the slower the emptying.
4. **Caloric content:** The duodenum regulates the delivery of calories.
5. **Particle size:** Large particles decrease the emptying rate.
6. **Intragastric pressure:** The greater the antral peristalsis and intragastric pressure, the faster the emptying.
7. **Pyloric sphincter resistance:** Greater resistance slows emptying, and vice versa.
8. **Duodenal pressure:** Increased duodenal pressure slows emptying, and vice versa.
9. **Negative feedback:** Control of emptying is mediated by neural and humoral factors activated by nutrients.

GASTRIC MOTOR DYSFUNCTION

- *The most common dysfunction is **gastroparesis,** which is delayed gastric emptying in the absence of mechanical obstruction.*
  - *–A long history of diabetes associated with peripheral neuropathy can cause diabetic gastroparesis.*
  - *–The failure to generate enough force to empty the stomach can be caused by a variety of disorders, such as abnormal slow-wave progression or loss of extrinsic innervation (eg, from vagotomy).*
  - *–The most common cause of delayed gastric emptying in adults is pyloric obstruction caused by scarring and edema from peptic ulcer disease.*
- *Disorders associated with **rapid gastric emptying** are often related to surgical procedures such as vagotomy or pyloric resection.*
  - *–Incompetence of the pyloric sphincter allows too rapid emptying of hypertonic material into the small intestine, resulting in dumping syndrome.*

–Vagotomy results in a loss of gastric compliance and an increased rate of emptying liquids.
–Patients with duodenal ulcers exhibit rapid gastric emptying, which may be due to a loss of duodenal negative feedback control mechanisms.

V. Gastric Secretion

A. The **gastric mucosa** has two main divisions: the oxyntic or parietal glandular mucosa, and the pyloric glandular mucosa.

B. The **oxyntic (parietal) glandular mucosa** comprises 85% of the total glandular region.

1. **Parietal cells** secrete hydrochloric acid and intrinsic factor (required for the intestinal absorption of vitamin B_{12}).
   a. A H^+-K^+ pump is responsible for parietal cell gastric acid secretion.
2. **Chief (peptic) cells** secrete multiple pepsinogens, which are converted to pepsins by low pH on the surface of the stomach and begin protein digestion.
   a. Chief cells are stimulated by both cAMP (secretin) and Ca^{2+} (ACh, gastrin, CCK) pathways.
3. **Enterochromaffin-like (ECL) cells** release histamine, which, along with acetylcholine and gastrin, stimulate parietal cells to secrete acid.
4. **Mucous cells** on the gastric gland surface secrete mucus that lubricates and protects the gastric mucosa through its high HCO_3^- content.
   a. Surface cells are stimulated to secrete mucus by acetylcholine, acids and prostaglandins.
5. **Mucous neck cells** secrete mucus and serve as stem cells for other glandular cells.

C. With increasing rates of gastric juice secretion, the [H^+] rises and the [Na^+] falls.

D. The **pyloric glandular mucosa** secretes mucus and GI regulatory peptides.

1. **Mucous cells** on the surface and glandular neck area secrete mucus that serves a protective role.
2. **G cells** secrete gastrin, a major stimulant of acid secretion and pepsinogen release, as well as mucosal growth.
3. **D cells** secrete somatostatin, a universal inhibitor peptide that inhibits gastric secretion.

E. The gastric mucosal barrier can be disrupted by various substances (Table 5–1).

1. Normal gastric mucosa is impermeable to H^+, thus preventing damage.

Table 5–1. Agents known to disrupt the gastric mucosal barrier.

Agent	Example
Weak acids	Aspirin
Alcohols	Ethanol
Nonsteroidal anti-inflammatory drugs	Indomethacin
Detergents	Bile salts

2. The permeability of this barrier is increased by salicylates, ethanol, and bile acids. As a result, acid diffuses back into the gastric mucosa, causing
   a. Pain due to stimulation of motility
   b. Acid-induced stimulation of pepsinogen secretion
   c. Acid-induced release of histamine that stimulates more acid secretion
   d. Increased capillary permeability and vasodilation (caused by locally released histamine), leading to edema of the mucosa
   e. Bleeding of dilated vessels, ranging from superficial to exsanguination
   f. In severe injury, inflammatory cells to release several agents such as leukotrienes, thromboxanes that produce ischemia and tissue damage.

F. There are three primary **stimulants of acid secretion** that act through Ca^{2+}/diacylglycerol or cAMP (Figure 5–4):
1. **Acetylcholine** released by diffuse efferent vagal fibers binds to muscarinic receptors on parietal cells.
2. **Gastrin** interacts with CCK_B receptors on parietal cells.
3. **Histamine** released from ECL cells in the fundus and from mast cells in the antrum binds to H_2 receptors on the parietal cells.
   a. Histamine potentiates the responses of the parietal cell to acetylcholine and gastrin. This interaction yields a response that is greater than the sum of the responses to each agent alone.
   b. This potentiation provides the basis for H_2-receptor blocking drugs (eg, cimetidine) that inhibit acid secretion.

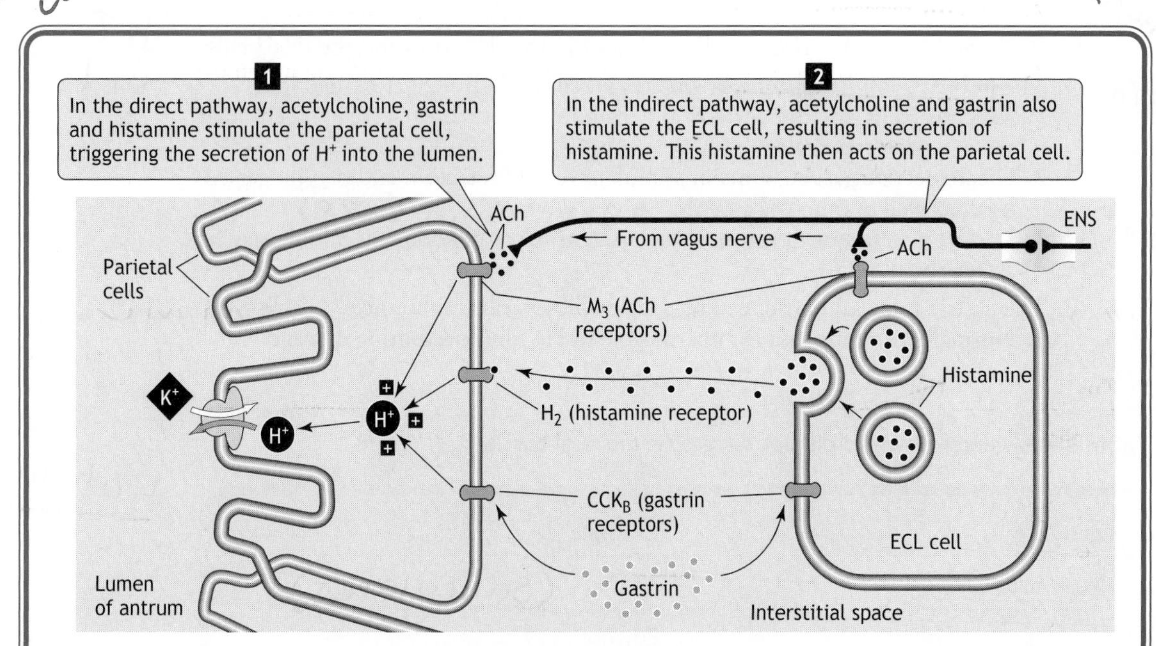

Figure 5–4. The direct and indirect actions of the three acid secretagogues: acetylcholine, gastrin, and histamine. ACh, acetylcholine; $CCKB_B$ cholecystokinin B; ECL, enterochromaffin-like; ENS, enteric nervous system. *(Reprinted from Boron and Boulpaep: Medical Physiology, Figure 41-5, © 2003 with permission from Elsevier, Inc.)*

Table 5–2. Mechanisms inhibiting gastric acid secretion.

Region	Stimulus	Mediation	Inhibits Gastrin Release	Directly Inhibits Acid Secretion
Antrum	Acid	Somatostatin	+	
Duodenum	Acid	Secretin	+	+
		Nervous reflex		+
	Hyperosmotic solutions	Unidentified enterogastrone		+
Duodenum and jejunum	Fatty acids	GIP	+	+

G. **The following mechanisms lead to secretory inhibition** (Table 5–2):
   1. **Somatostatin** released by gastric antral D cells when luminal pH falls below 2.0 inhibits further gastrin release and acts as the primary mechanism of inhibition of acid secretion.
   2. **Secretin** released into the circulation from S cells in the duodenum acts on parietal cells to inhibit acid secretion.
   3. **Prostaglandin E_2** is thought to inhibit acid secretion by blocking histamine stimulation of parietal cells.

H. The three phases of gastric secretion associated with eating are cephalic, gastric, and intestinal (Table 5–3).

Table 5–3. Phases of gastric secretion.

Phase	Stimulant	Pathway	Mediator	% of Total Secretion
Cephalic	Sight, smell, and taste of food	Direct vagovagal —gastrin-releasing peptide	Acetylcholine	> 30
Gastric	• Distention • Amino acids • Protein digestion products	Vagovagal intramural G-cell stimulation	Gastrin	> 50
Intestinal	• Distention • Protein digestion products	Amino acid in blood	Gastrin	5–10

GASTRIC SECRETORY DYSFUNCTION

- ***Hypersecretion:*** *associated pathophysiology*
  *–**Duodenal ulcer** is associated with Helicobacter pylori infection that leads to increased gastric acid secretion. Acid hypersecretion causes metaplasia of gastric cells in the duodenum that are colonized by H pylori, leading to duodenal ulcer formation. Plasma gastrin levels are normal or only slightly elevated.*
  *–**Zollinger-Ellison syndrome** (gastrinoma) involves a gastrin-secreting tumor in the pancreas or intestine, which produces elevated levels of circulating gastrin, leading to a high level of gastric acid secretion and resulting in peptic ulceration.*
- ***Hyposecretion:*** *associated pathophysiology*
  *–In **gastric ulcer disease,** the reflux of bile and pancreatic enzymes from the duodenum causes gastric ulceration.*
  *–In **pernicious anemia,** the lack of intrinsic factor secretion causes vitamin B_{12} deficiency that leads to failure of red blood cell maturation and microcytic anemia.*
  *–This condition is often associated with **gastric atrophy, achlorhydria,** and high **gastrin levels** often seen in the elderly. Thus, intrinsic factor secretion by parietal cells makes the stomach essential for life.*

VI. Motility of the Small Intestine

A. The **small intestine** is the **major site for digestion and absorption** of food and is divided into three sections: the duodenum, jejunum, and ileum (Figure 5–5).

1. Ninety-five percent of nutrients are usually absorbed by the time a meal reaches the distal jejunum.
2. The remainder of the intestine is devoted primarily to absorption of water and electrolytes.
3. The entire small intestine has the capacity for absorption of nutrients, which provides a functional reserve for the body.
4. The ileum has specific absorptive mechanisms for cobalamin (vitamin B_{12}) and bile acids.
5. Transit time through the small intestine is 2–4 hours for chyme.

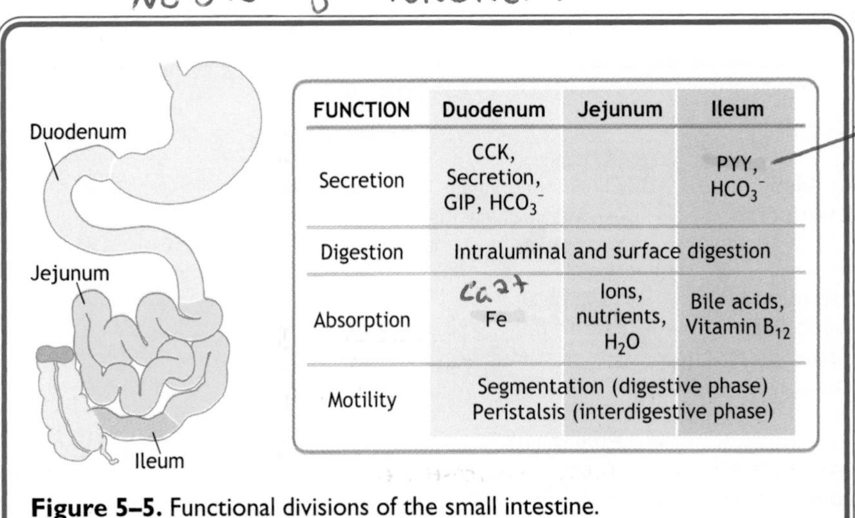

FUNCTION	Duodenum	Jejunum	Ileum
Secretion	CCK, Secretion, GIP, HCO_3^-		PYY, HCO_3^-
Digestion	Intraluminal and surface digestion		
Absorption	Fe	Ions, nutrients, H_2O	Bile acids, Vitamin B_{12}
Motility	Segmentation (digestive phase) Peristalsis (interdigestive phase)		

Figure 5–5. Functional divisions of the small intestine.

B. Digestion and absorption of food depend on normal **contractile behavior of the small intestine.**

1. Intestinal **slow waves** determine the frequency and patterns of contractions (Figure 5–6).
2. The frequency of slow waves is highest in the proximal small intestine (12/min).
3. There is a stepwise decrease in frequency from the duodenum (12/min) to the ileum (8/min).
4. Motility pattern differs in the Fed and Fasted states.
5. Fed motor activities associated with contractions are **segmentation** (mixing) and **peristalsis** (propulsion) (Figure 5–7).
   a. In the small intestine, contraction of circular muscle results from a temporary removal of the inhibitory effects of the enteric nervous system.
   b. The timing of contractions is determined by **slow wave depolarization.**
   c. **Segmentation** is characterized by isolated contractions, which moves chyme in both directions and is the most common type of intestinal contraction.
   d. Segmentation increases the exposure of chyme to enzymes and contact with absorbing cells.
   e. Peristalsis is not considered to be an important component of intestinal transit because it moves chyme only a few centimeters at a time.
6. In **fasting motor activity,** the purpose of MMCs is to keep the small intestine swept clean of bacteria, undigestible meal residua, desquamated cells, and secretions.
   a. Inhibition of intestinal motor activity in the rat (with morphine) leads to bacterial overgrowth in the ileum within 6 hours.
   b. When a segment of the intestine is severed, it generates spontaneous MMCs at a rate higher than in the intact intestine.
   c. Not every MMC progresses all the way to the terminal ileum.

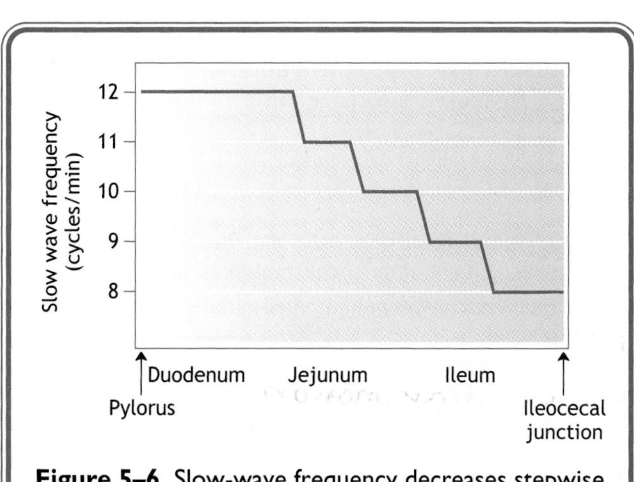

Figure 5–6. Slow-wave frequency decreases stepwise from the duodenum to the ileum.

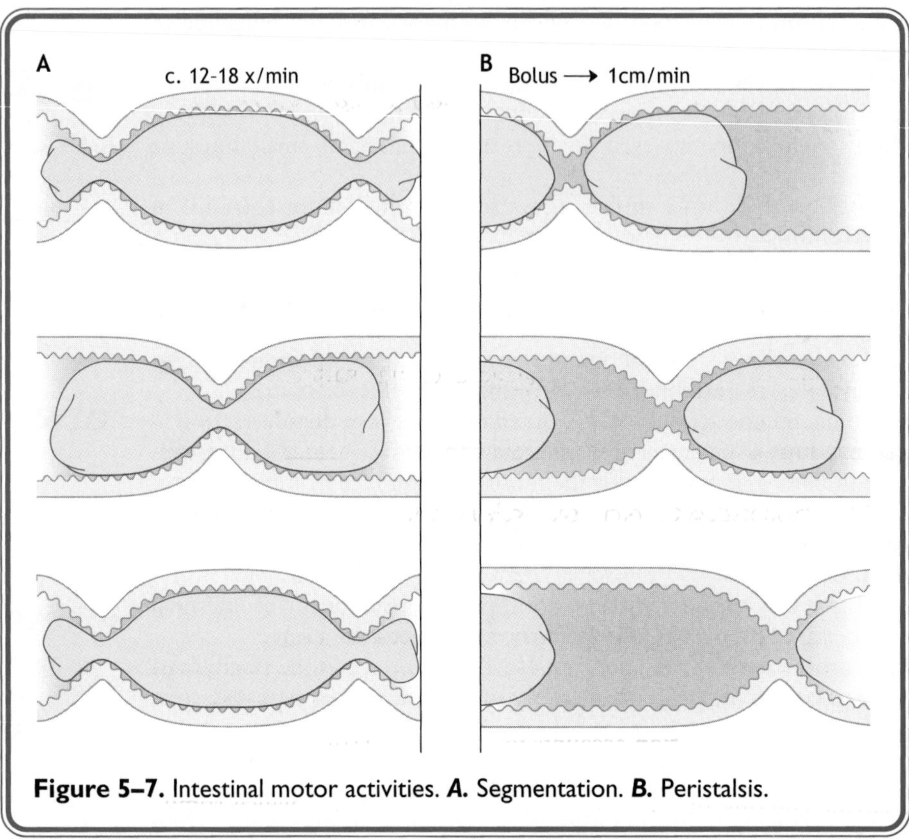

Figure 5–7. Intestinal motor activities. ***A.*** Segmentation. ***B.*** Peristalsis.

d. Feeding interrupts the interdigestive MMC and initiates the fed pattern of motility, which is more conducive to absorption than the fasting pattern.

e. Although the physiologic mechanisms responsible for switching from the fasting to the fed motor pattern are not known, infusion of neurotensin, a GI peptide released with feeding, is associated with inhibition of MMCs in humans.

INTESTINAL MOTOR DYSFUNCTION

- ***Symptoms such as nausea, vomiting, abdominal distension, colic, diarrhea,*** *and* ***constipation*** *may result from abnormalities in moving luminal contents through the small intestine.*
- ***Vomiting*** *is a complex, coordinated set of motor discharges programmed in the medullary vomiting center.*
  - *–Vomiting is initiated by direct activation of the vomiting center or by activation of the medullary chemoreceptor trigger zone.*
  - *–Prior to vomiting, intense spike activity appears in the mid small intestine and travels up to the pylorus at the rate of 2–3 cm/s.*
  - *–The stomach and esophagus are relaxed.*
  - *–Gastric contents are then moved up to and out of the mouth by forceful contraction of abdominal muscles (retching) and the diaphragm.*

–Blood-borne chemicals, such as apomorphine, stimulate vomiting through the chemoreceptor trigger zone.
–Afferents from the stomach and intestine can stimulate the vomiting center directly.
–Projectile vomiting, which is forceful emesis not associated with nausea, is caused by direct stimulation of the medullary vomiting center.

- ***Peristaltic rush** is abnormal in humans but common in animals that consume feces.*
  –It is characterized by strong peristaltic waves moving chyme large distances.
  –It results in maldigestion, malabsorption, and diarrhea in humans.

VII. Exocrine Pancreas

A. The pancreas has a dual function, with 90% exocrine cells and 10% endocrine mass.

B. The pancreatic duct runs the length of the gland and joins with the common bile duct before opening into the duodenum at the ampulla of Vater.

C. The **functional unit** of the exocrine pancreas **consists of acinar and ductal cells.**

1. **Acinar cells** are specialized protein synthesizing cells that produce enzymes, and **ductal cells** are specialized for fluid and electrolyte transport and generate a watery HCO_3^- secretion to neutralize gastric acid entering the duodenum.
2. The amount of HCO_3^- secretion is proportional to the load of gastric acid delivered to the duodenum below the threshold pH of 4.5.

D. During the cephalic and gastric phases of digestion, some pancreatic secretion occurs as a result of vagovagal cholinergic reflexes and increased serum gastrin.

E. The **intestinal phase of digestion** accounts for most of the stimulation of pancreatic secretion via secretin and CCK.

1. Acidic chyme entering the duodenum causes secretin release, which stimulates an isotonic Na HCO_3 secretion from ductal cells.
2. The HCO_3^- in pancreatic ductal secretions neutralizes acid, thus removing the stimulus for further secretion of secretin.

F. **Fat and protein digestive products** entering the duodenum **stimulate CCK release,** which stimulates **pancreatic enzyme secretion.**

1. These enzymes hydrolyze proteins, starches, and fats.
2. About 80% of pancreatic enzymes produced are proteolytic and are released in their inactive, or pro, form.
3. Trypsinogen is converted to trypsin by **enterokinase,** a brush cell border enzyme. Trypsin then converts the remaining proteolytic proenzymes to their active forms.
4. The total amount of the enzymes produced is secreted. There is no enzyme storage in the pancreas.
5. Ca^{2+} is the major second messenger for pancreatic enzyme secretions.

G. Secretin and CCK potentiate the stimulatory effects of one another.

1. In addition, acetylcholine, from parasympathetic innervation to the pancreas, potentiates the effects of CCK and secretin. Thus, vagotomy may decrease the pancreatic secretory response to a meal by more than 50%.
2. CCK may also act through a vagovagal pathway to reflexly stimulate pancreatic secretion.
3. Several mechanisms protect the pancreas from autodigestion by proteolytic enzymes

a. enzymes are kept inside secretory granules while in pancreas
b. converting enzyme is located away from pancreas
c. trypsin inhibitor prevents trypsin activation inside pancreas

CHRONIC PANCREATITIS

CLINICAL CORRELATION

- *Chronic pancreatitis is most often associated with a history of chronic alcohol abuse.*
- *Malabsorption does not occur until the pancreatic enzyme secretory capacity is reduced by 90%.*
- *Decreased secretion of digestive enzymes in chronic pancreatitis results in fat maldigestion, causing a major calorie loss and malabsorption leading to decreased vitamin B_{12} absorption.*
- ***Treatment** involves oral **administration of pancreatic enzymes.***

VIII. Biliary Secretion

A. **Bile** is secreted continuously by the canaliculi of the liver, eventually flows into the duodenum via the common bile duct.
   1. The secretion rate depends on whether a fed or fasting state exists.
   2. The gallbladder is a storage reservoir that can deliver bile to the duodenum for the solubilization of dietary lipid.

B. Bile contains bile salts, lecithin (a phospholipid), cholesterol, bile pigments (eg, bilirubin), water, and electrolytes.

C. Bile constituents are dissolved in an alkaline solution resembling pancreatic juice. Bile plays an important role in the intestinal digestion and absorption of lipids.

D. **Primary bile acids**—cholic acid and chenodeoxycholic acid—are synthesized by the liver from cholesterol. The lipid-soluble bile acids are conjugated with either glycine or taurine.

E. Because they are ionized at neutral pH, **conjugated bile acids** exist as salts of sodium or potassium and, therefore, **are known as bile salts.**

F. **Secondary bile acids** are formed by deconjugation and dehydroxylation of the primary bile salts by intestinal bacteria, forming deoxycholic acid from cholic acid and lithocholic acid from chenodeoxycholic acid.

G. Lithocholic acid is hepatotoxic and is normally excreted in feces.

H. The bile acid pool, which under normal conditions is constant in size (about 2–4 g), is a mass of primary and secondary bile acids.

I. **Bile acid absorption** occurs largely in the ileum, where an active transport mechanism exists. Approximately 95% of the total pool is absorbed.

J. Colonic absorption of bile acids is minimal. In excess, bile acids can cause a concentration-dependent increased secretion in the colon, leading to watery diarrhea when in excess.

K. Bile salts regulate their own synthesis by negative feedback from the intestine.

L. Bile acid synthesis is increased with decreased return of bile acids to the liver and is decreased with increased return of bile acids.

M. This recycling of bile salts to the liver via the portal circulation after reabsorption by the intestine is called the **enterohepatic circulation** of bile salts (Figure 5–8). Bile acids are taken up by hepatocytes from the blood, reconjugated, and then resecreted into bile. Bile acids must be recirculated 3–5 times for digestion of a normal meal.

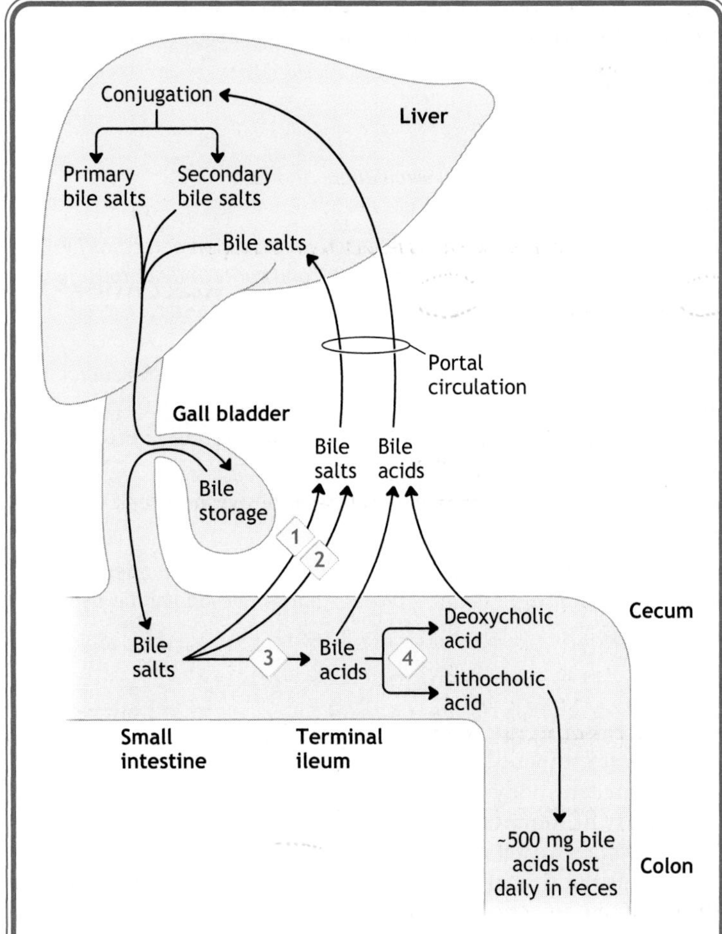

Figure 5–8. Enterohepatic circulation. ***1*** and **2** represent bile salts of hepatic origin that are passively absorbed into the portal circulation, whereas ***3*** and ***4*** represent bile acids in the intestinal lumen that are acted on by bacteria and are dehydroxylated to secondary bile acids (eg, deoxycholic acid), which is actively absorbed in the ileum, and lithocholic acid, which is primarily excreted in feces.

N. **Bile secretion** is regulated primarily by meal-stimulated CCK, which causes gallbladder contraction and sphincter of Oddi relaxation.

O. Secretin stimulates bile ductules and ducts to secrete a watery HCO_3^- rich fluid.

P. When bile salts become concentrated they form **micelles,** or large molecular aggregates that are water soluble on the outside and lipid soluble on the inside. Thus, they provide a vehicle for transport of lipid-soluble materials in the aqueous medium of the small intestine.

Q. Micelles are vital for fat digestion and absorption. Damage or removal of the distal ileum causes bile salt deficiency and leads to fat maldigestion and malabsorption.

CHOLELITHIASIS (GALLSTONES)

CLINICAL CORRELATION

- *Cholelithiasis is a disturbance in bile secretion and cholesterol elimination.*
- *Most gallstones are cholesterol stones.*
- *Epidemiologic factors associated with gallstone formation include geographic location and ethnicity (desert Native Americans), age (40+), gender (primarily female), obesity (+), and parity (multiparous).*
- *Bile is the only route for excretion of cholesterol in the body. When the amount of cholesterol exceeds the concentration of bile salts to solubilize it, cholesterol stones may precipitate out.*
- *Thus, **lithogenic,** or stone-forming, bile is supersaturated with cholesterol resulting from high-cholesterol **production or low–bile acid production.***
- *Gallstones often block bile ducts, thereby preventing bile from entering the intestine and causing severe pain in the upper-right quadrant.*
- *Bile stasis results in sequestration of bile in the gallbladder and a blunted secretory response to CCK.*
- *Abdominal ultrasonography is the primary means of diagnosis of gallstones.*
- ***Primary treatment** is surgery via laparoscopic cholecystectomy. A new nonsurgical treatment involves administration of synthetic bile salts to dissolve stones, a process that can be long and expensive.*

IX. Digestion and Absorption

A. Small Intestine: Nutrient Entrance to the Body

1. All nutrients (carbohydrate, protein, fat, vitamins, and minerals) and most fluids and electrolytes enter the body through the small intestine.
2. The **surface area consists of mucosal folds, villi,** and **microvilli,** which together occupy a 2,000,000 cm^2 total area.
3. **Absorption** takes place at the **tips of the villi; secretion** occurs in the **crypt region of the villi.**
4. The crypt is the birthplace of new mucosal cells, and new cells migrate up the lateral surface toward the tip of the villus.
5. The total life span of mucosal cells is 4–5 days, after which they are sloughed off into the lumen.

B. Carbohydrate Digestion and Absorption

1. Carbohydrates contribute more than 50% of caloric intake, and starches are the predominant type.
2. **Starch digestion** begins in the mouth via **salivary amylase,** or **ptyalin.** Salivary amylase activity is partially inhibited by gastric acid (Figure 5–9).
   a. **Pancreatic amylase** hydrolyzes most starch to disaccharides, trisaccharides, and α-limit dextrins and is essential for starch digestion.
   b. **Brush cell border disaccharidase** activity results in the monosaccharides glucose, galactose, and fructose. For example, **sucrase** breaks down sucrose to glucose and fructose; **lactase** converts lactose to glucose and galactose.
3. **Glucose and galactose** are **absorbed** (via secondary active transport) **through** a **sodium-dependent cotransporter** known as **SGLT 1.** A high luminal concentration of sodium facilitates absorption and vice versa.

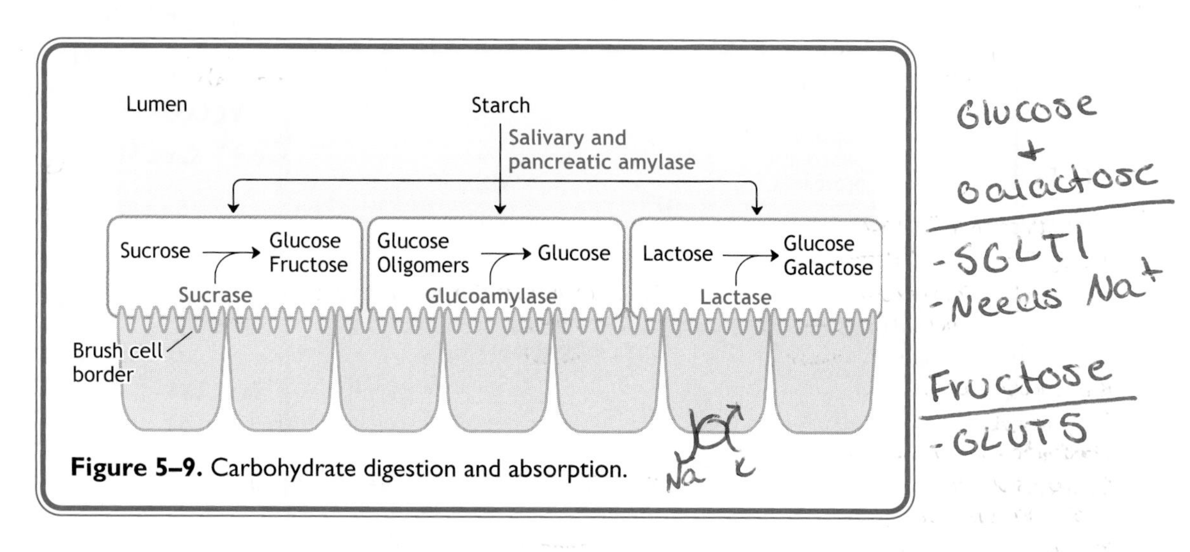

Figure 5–9. Carbohydrate digestion and absorption.

4. **Fructose** enters by **facilitated diffusion via** glucose transporter 5 (**GLUT 5**) that does not require sodium.
5. All monosaccharides are transported out of the enterocyte into capillaries by **GLUT 2.**
6. Except for lactase, the levels of disaccharidases are adaptable to the diet.
7. Normally all carbohydrates have been absorbed by the time the chyme reaches the mid-jejunum.

LACTASE DEFICIENCY

CLINICAL CORRELATION

- *This deficiency occurs in 70% of nonwhites and causes lactose sensitivity leading to bloating, gas, and diarrhea when milk sugar or lactose is consumed.*
- *Symptoms depend on the lactose load, lactase presence, transit time, and the ability of colonic bacteria to metabolize lactose to H_2, CO_2 and short-chain fatty acids.*
- *Milk consumption normally results in a 25 mg/dL increase in plasma glucose. Those with lactase deficiency exhibit less than a 20 mg/dL rise and exhibit symptoms of bloating, cramps, diarrhea, and an increase in breath H_2.*
- *Synthetic lactase can be administered orally to lactase-deficient persons to prevent symptoms.*

C. Protein Digestion and Absorption

1. Protein digestion is **initiated** in the stomach **by the action of pepsin** and requires hydrolysis to oligopeptides or amino acids before absorption. (Figure 5–10)
2. Most **protein digestion takes place in the small intestine** by pancreatic proteases.
   a. Specific proteolytic enzymes split peptides and oligopeptides into amino acids.
3. Peptides are also broken down by brush cell border peptidases into amino acids.
4. Small peptides may be absorbed intact into the enterocyte and intracellular peptidases hydrolyze them into free amino acids that enter the circulation.

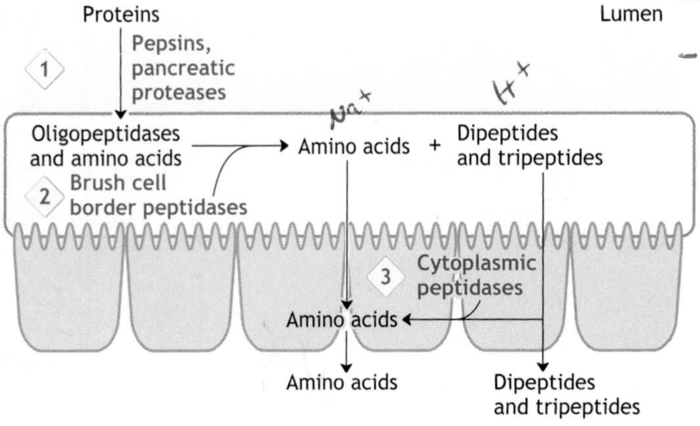

Figure 5–10. Protein digestion occurs (*1*) in the stomach and intestinal lumen, (**2**) in the intestinal brush cell border, and (**3**) intracellularly.

5. Absorption of whole protein occurs primarily during the neonatal period.
6. Free luminal amino acids are absorbed via sodium-dependent secondary active transport.
7. There are several different sodium-dependent carrier systems for different classes of amino acids.
8. The apical absorption of oligopeptide is via an H^+/oligopeptide cotransporter.
9. Some amino acids are more readily absorbed as peptides than as free amino acids.
10. Most amino acids and peptides are absorbed in the jejunum.

GLUTEN ENTEROPATHY

CLINICAL CORRELATION

- *This syndrome results from a hypersensitivity to dietary gluten protein and is also known as* ***celiac sprue*** *or* ***gluten-sensitive enteropathy.***
- *It leads to flattening of microvilli and generalized malabsorption.*
- *Normal function returns if an affected person adheres to a gluten-free diet.*

D. Fat Digestion and Absorption

1. **Triglycerides** are the **most abundant lipids** in the diet.
2. Special mechanisms are present to digest and absorb fats because they are insoluble in water.
3. **Fat digestion begins in the stomach,** where fats are emulsified; about 30% of fats are digested by lingual and gastric lipases.
4. **Most digestion and absorption of lipids** occurs **in the small intestine,** where bile micelles emulsify fat and pancreatic lipases digest fat.
5. The major products of triglyceride digestion are 2-monoglycerides and **free fatty acids (FFAs)** (Figure 5–11).

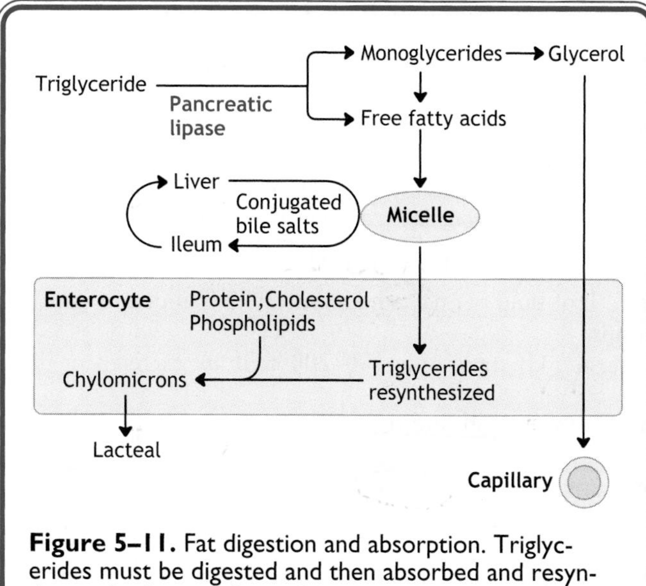

Figure 5–11. Fat digestion and absorption. Triglycerides must be digested and then absorbed and resynthesized, whereas glycerol does not have to be digested and can move directly to the capillaries. Chylomicrons are large fatty droplets that are too large for capillaries and so must enter lymph ducts (lacteals).

6. Adherence of pancreatic lipase to the emulsion requires **colipase,** a polypeptide secreted by the pancreas that allows lipase to bind and hydrolyze triglycerides to 2-monoglycerides and FFAs.
7. The products of fat digestion are solubilized by incorporation into mixed micelles composed of bile salts, monoglycerides, FFAs, phospholipids, cholesterol, and fat-soluble vitamins.
8. Micelles diffuse through the unstirred layer and cross the enterocyte brush cell border of the intestine.
9. The digested lipids are released from the micelles and then diffuse into the mucosal cells. The bile salts are later reabsorbed from the ileum by sodium-dependent secondary active transport.
10. Once inside the enterocyte, absorbed monoglycerides and FFAs are resynthesized into triglycerides and along with cholesterol esters, assembled into fatty droplets known as **chylomicrons.**
11. After a layer of protein and phospholipid is added, the chylomicrons pass through the basolateral membrane and enter the lymphatic circulation during feeding.
    a. During fasting the intestine secretes very-low-density lipoproteins.
12. Short- and medium-chain fatty acids are more water soluble and can pass by simple diffusion into the portal circulation without digestion.

TION

- *carbohydrate or protein malabsorption, fat malabsorption usually results from cy or bile salt deficiency.*
- ***ea,** or fatty, bulky, foul-smelling stools that lead to significant caloric and fat-soluble vitamin deficiencies.*
- *Administration of synthetic short- and medium-chain fatty acids that do not require digestion can alleviate the caloric and vitamin deficiencies that result from pancreatic and biliary insufficiency.*

E. Fluid and Electrolyte Absorption

1. Adults take in about 2 L of fluid per day, and another 7 L is added to the GI tract through secretions.
2. Of the 9 L/d entering the GI tract, only 100–200 cc/d are excreted in the stool.
3. **Most fluid is absorbed in the small intestine,** even though both the small and large intestine absorb fluids and water absorption is most efficient in the colon.
4. All water reabsorption in the gut is passive and secondary to solute movement. Solutes can be electrolytes such as sodium or nonelectrolytes such as glucose.
5. The water absorbed in the gut is later available for secretions that are added to the next meal as well as to replace fluids lost through urination, perspiration, and respiration.
6. Sodium is transported from the lumen into the lateral intercellular space where an osmotic gradient is created causing water to flow into the intercellular space. The water flow increases hydrostatic pressure in the intercellular space, which causes fluid flow into the interstitial space and blood.
7. Na/Glucose and Na/amino acid cotransport from the lumen into enterocytes stimulates water absorption.
8. Na^+ is absorbed by various mechanisms: passive diffusion via a Na^+ channel (in the colon), cotransport with solutes, cotransport with Cl^-, and exchange with H^+.
9. Once in the intestinal cell, Na^+ is actively transported across the basolateral membrane by Na^+/K^+-ATPase.
10. Parallel Na-H and Cl-HCO3 exchange in the ileum and proximal colon is the primary mechanism of Na+ and Cl- absorption during the interdigestive period.
11. Epithelial Na^+ channels are the primary mechanism of Na^+ absorption in the distal colon.
12. Cl^- secretion occurs in the crypts of both the small and large intestine.
13. Most K^+ is absorbed passively via solvent drag except for active absorption in the rectum by an active H-K pump.
14. K^+ is actively secreted in the colon and rectum in response to aldsterone and cAMP. Thus, chronic diarrhea can lead to significant hypokalemia.
15. Thus, the small intestine secretes HCO_3^- while the colon secretes both HCO_3^- and K^+.

CHOLERA

- *Cholera epidemics continue to be a **major cause of fatalities worldwide.***
- *Cholera toxin irreversibly stimulates the cAMP-dependent Cl^- pump in intestinal cells resulting in massive Cl^--rich watery diarrhea.*

- *Death is caused by extreme dehydration and electrolyte imbalance.*
- ***Treatment** involves hydration and electrolyte replacement, which can now be done through formulations similar to sport drinks that promote rapid water and electrolyte absorption. These solutions contain glucose and fructose plus Na^+ and K^+, which establish an osmotic gradient in enterocytes and a hydrostatic gradient to quickly move fluid and electrolytes into blood.*

F. Vitamin B_{12} (cobalamin) Absorption

1. Cobalamin (CBL) is bound to proteins in food
2. Gastric acid and pepsin release CBL from dietary protein
3. Salivary and gastric glands secrete haptocorrin which binds to CBL
4. Gastric parietal cells secrete intrinsic factor (IF)
5. Pancreatic proteolytic enzymes degrade haptocorrin binding to CBL
6. The IF-CBL complex forms
7. Ileal cells absorbs IF-CBL complex
8. Atrophy of the gastric mucosa results in a deficiency of IF and acid secretion that causes **CBL deficiency**
9. Vitamin B_{12} deficiency produces pernicious anemia which is characterized by glossitis and neuropathy
10. Parenteral administration of **CBL** reverses and prevents the manifestations of pernicious anemia.

G. Calcium and Iron Absorption

1. **Active Ca^{2+} absorption** occurs **in the duodenum** and is **primarily regulated by vitamin D_3,** or **1,25-dihydroxycholecalciferol,** which stimulates synthesis of Ca^{2+}-binding protein in enterocytes.
2. Ca^{2+} is absorbed by passive diffusion throughout the small intestine and is regulated by vitamin D.
3. The active transport of Ca^{2+} across the basolateral membrane by Ca^{2+}-pump and a Na-Ca exchanger is also regulated by vitamin D_3.
4. **Parathyroid hormone** activates vitamin D to vitamin D_3 in the kidney, resulting in more conversion when blood Ca^{2+} levels are low.
5. Fat malabsorption due to pancreatic or bile deficiency leads to decreased vitamin D absorption, which subsequently decreases Ca^{2+} absorption.
6. **Iron absorption** occurs primarily in the duodenum and is tightly regulated based on the body's need.
7. Heme and nonheme iron are absorbed by distinct cellular mechanisms.
   a. Heme iron is absorbed more efficiently than nonheme iron.
8. **Non-heme** iron binds to a divalent cation transporter on the brush cell border membrane which cotransports Fe^{2+} and H^+ into the cell.
9. Heme iron from myoglobin and hemoglobin is absorbed by endocytosis or a brush border protein and is then split by heme oxidase to release free Fe^{3+} which is converted to Fe^{2+} and then handled the same as nonheme iron.
10. If the need is great, iron is transferred rapidly into the blood to be carried by a protein called **transferrin.**
11. If the need is low, iron is bound to apoferritin in the cell to form **ferritin**, the storage form of iron.
12. After hemorrhage, it takes 4–5 days before more iron is absorbed because the iron-loaded intestinal cells must be sloughed off and new cells must migrate to the tips of the villi to adjust to the new need for iron.

13. Ascorbic acid reduces nonheme iron from the ferric (Fe^{3+}) to the ferrous state (Fe^{2+}) to increase absorption while tannins in tea form insoluble iron complexes and reduce absorption.

X. Motility of the Colon and Rectum

A. The colon conserves water and electrolytes and is involved in the formation, storage, and periodic elimination of indigestible materials.
   1. **Haustra,** or colonic sacculations, **result from** the **anatomic arrangement of the longitudinal muscle,** which is **concentrated in three bundles,** or **teniae coli,** instead of a solid sheath in the upper GI tract.
   2. There are no haustra in the lower colon or rectum because the longitudinal muscle forms a uniform coat again.
   3. Whereas the transit of food through the stomach and small intestine is measured in hours, food transit through the colon is measured in days.
   4. The majority of mixing and delay in transit occurs in the right colon.
   5. The frequency of slow waves in the colon increases from proximal to distal in contrast to that of the small bowel.

B. **Haustral segmentation contractions** occur 90% of the time, shuttling contents slowly back and forth to enhance absorption of water and electrolytes.
   1. **Multihaustral segmentation,** in which several haustra contract together and move contents a short distance, occurs 10% of the time.
   2. Mass movements occur usually after the first meal of the day and move material a long distance, often stimulating the urge to defecate. Mass movements are associated with the **gastrocolic reflex,** initiated by distention of the stomach with food, or the **orthocolic reflex,** stimulated by standing after reclining overnight.

C. **Defecation urge** is stimulated by distention of the rectal sigmoid area, which elicits the **rectosphincteric reflex,** or relaxation of the internal anal sphincter, and voluntary contraction of the external anal sphincter.
   1. If defecation is not socially appropriate, both the external and internal anal sphincter contract and an **adaptive relaxation reflex,** or decreased sensitivity to rectal wall distention, occurs.
   2. Mechanoreceptors in the rectal wall can discriminate between solid, liquid, or gas. This ability is lost in persons with ulcerative colitis due to mucosal damage.
   3. If defecation is socially appropriate, the internal and external anal sphincters relax and a **Valsalva maneuver,** or forced expiration against a closed glottis, is performed.

D. The **contractile activity of the colon** is **inhibited by the enteric nervous system** as in the small intestine.
   1. Stimulation of parasympathetic innervation increases colonic contraction.
   2. Stimulation of sympathetic nerves to the colon suppress motility.
   3. Fatty chyme in the ileum or colon releases **peptide YY,** which **inhibits** colonic and gastric **motility** and gastric and pancreatic **secretions.**

COLONIC DYSFUNCTION

- ***Diarrhea*** *occurs when the* ***volume of fluid*** *delivered to the colon* ***exceeds its absorptive capacity,*** *resulting in stool water content greater than 500 cc.*

Fiber:
Insoluble = apple skins, nuts, popcorn, wheat bran
Soluble = oat bran, citrus fruit, bean

–Diarrhea is caused by decreased absorption of fluid and electrolytes or increased secretion of fluid and electrolytes.
–Antidiarrheal drugs work either to increase fluid absorption or decrease secretion.

- ***Hirschsprung disease*** *(**aganglionosis,** or **megacolon**) is a congenital **absence of the enteric plexus** in the distal colon.*
  –With no inhibitory neurons present, colonic tone is increased, resulting in prolonged constipation.
  *–The area above the contracted segment becomes grossly dilated, causing **megacolon.***
  –Treatment involves removal of the tonically contracted area and reattachment to normal segments.
  –People with achalasia often exhibit megacolon.

CLINICAL PROBLEMS

A 25-year-old woman with a history of type 1 diabetes mellitus (ie, insulin deficient) since age 15 complains of prolonged constipation, abdominal distention, and severe heartburn. A barium GI and small bowel examination demonstrates a dilated stomach but no evidence of gastric outlet obstruction. The colon is filled with feces.

1. Which of the following statements about this case is correct?
   A. The symptoms are due to congenital absence of inhibitory neurons in the stomach.
   B. The treatment of choice is a high-fiber diet supplemented with laxatives.
   C. The symptoms are due to delayed gastric emptying associated with diabetic neuropathy.
   D. The diagnostic evidence points to peptic ulcer disease.
   E. Gastric surgery is the best choice to relieve the symptoms.

2. Which of the following slows gastric emptying?
   A. Fat in the duodenum
   B. Starch in the duodenum
   C. Protein in the duodenum
   D. High pH in duodenal chyme
   E. Isotonic NaCl in the duodenum

A 50-year-old painter gives a history of severe epigastric pain; tiredness; and oily, foul-smelling diarrhea. Although his appetite has been good and he has been eating a well-balanced diet, he has lost 20 lb over the past 5 months. He admits to weekend abuse of alcohol for over 20 years. On admission to the hospital, his serum amylase (25 IU/L; normal 30–110 IU/L) and lipase levels (20 IU/L; normal 23–100 IU/L) were decreased, and stool analysis showed a triglyceride level of 18 g (normal < 7 g) and undigested meat fibers. Serum bilirubin levels were normal, and no evidence of jaundice was seen. An abdominal x-ray showed many sites of calcium salt deposits in the pancreas.

3. What do the fat levels and undigested meat fibers in the stool suggest?
   A. Gastric atrophy
   B. Pancreatic exocrine insufficiency

C. Vitamin B_{12} deficiency
D. Peptic ulcer disease
E. VIP-secreting tumor

A 60-year-old man has a 2-month history of dysphagia, or swallowing difficulties. Barium swallow studies reveal a stricture at the gastroesophageal junction. Esophageal manometric studies showed an increased resting gastroesophageal sphincter pressure and a failure of the sphincter to relax with swallowing. In addition, there is an absence of progressive peristaltic contractions after swallowing.

4. What diagnosis do these results suggest?
   A. GERD
   B. Achalasia
   C. Hirschsprung disease
   D. Secondary peristaltic waves
   E. Water brash

A 65-year-old woman with a long history of Crohn's disease is admitted to the hospital with severe weight loss, general debility, and a complaint of severe watery diarrhea. She has had multiple bowel resections, with the most recent being a removal of 150 cm of ileum. Current therapy consists of vitamin B_{12} and folic acid supplements. A contrast radiographic study of the surgical area shows no evidence of recurrent disease. Stool study results are negative for mucus and blood.

5. What is the most likely cause of her diarrhea?
   A. Folic acid deficiency
   B. Bile acid malabsorption
   C. VIP-secreting tumor
   D. Cholera toxin
   E. Recurrent Crohn's disease

A 63-year-old woman has undergone a total gastrectomy for gastric carcinoma. After surgery, she received no nutritional supplementation or counseling. Four years later she appears at her physician's office severely anemic and extremely fatigued.

6. What is the most likely cause of her anemia?
   A. Vitamin D deficiency
   B. Vitamin K deficiency
   C. Vitamin B_{12} deficiency
   D. Vitamin A deficiency
   E. Vitamin E deficiency

7. The mucopolysaccharide, or mucopolypeptide in the normal stomach secretion, that combines with vitamin B_{12} and makes it available for absorption by the gut is called
   A. Secretin
   B. Intrinsic factor

C. Pancreozymin
D. Antihemophilic factor A
E. Pyridoxine

ANSWERS

1. C is correct. The delayed emptying of solids and liquids from the stomach (gastroparesis) occurs in 30–50% of patients with diabetes. The phenomenon is thought to be due to vagal autonomic neuropathy to the stomach. Distention of the caudad stomach stimulates an excitatory vagovagal reflex that stimulates mixing and then emptying activity. With congenital absence of inhibitory neurons (choice A) there would be tonic contraction of gastric musculature with little filling or emptying. Treatment with a high-fiber diet (choice B) would increase intestinal motility but would not help poor gastric motility. Diagnostic evidence in this case does not point to peptic ulcer disease, as there is no report of excess acid secretion (choice D). Surgical removal of gastric tissue (choice E) would not improve the symptoms. Optimal treatment includes optimal glycemic control, a low-residue diet, and prokinetic agents that stimulate gastric motility.

2. A is correct. Increasing the fat content of the duodenum stimulates the release of inhibitory neural (enterogastric reflex) and hormonal cholecystokinin (CCK) feedback mechanisms, which reduce gastric motility. Starch in the duodenum (choice B), protein in the duodenum (choice C), high pH in the duodenum (choice D), and isotonic NaCl in the duodenum (choice E) have little influence on gastric emptying.

3. B is correct. The high levels of fat and undigested meat fibers in the stool and low enzyme level indicate maldigestion and a deficiency in pancreatic enzyme secretion. Atrophy of gastric mucosa (choice A) is not usually associated with severe maldigestion. Vitamin B_{12} deficiency (choice C) leads to pernicious anemia, not to pancreatic enzyme deficiency. Peptic ulcer disease (choice D) is associated with increased gastric acid secretion, not pancreatic enzyme deficiency. VIP-secreting tumor (choice E) is associated with a severe watery diarrhea, not the oily, foul-smelling diarrhea described in this case.

4. B is correct. Increased lower esophageal sphincter pressure and the absence of esophageal peristalsis are characteristics of achalasia (no relaxation of esophagus). The condition is due to an absence of inhibitory intramural neurons in the esophagus. Gastroesophageal reflux disease (GERD) (choice A) is associated with decreased lower esophageal pressure, not increased lower esophageal pressure. Hirschsprung disease (choice C) or megacolon is caused by the absence of inhibitory neurons in the wall of the distal colon causing contraction of the affected segment and prolonged constipation. Secondary peristaltic waves (choice D) in the esophagus are clearing waves that remove residual material remaining after a primary peristaltic wave is complete. Water brash (choice E) is a sudden increase in flow of saliva thought to be produced by a reflex due to refluxed gastric acid into the distal esophagus.

5. B is correct. The terminal ileum contains specialized cells responsible for the absorption of bile salts by active transport. Bile salts are necessary for adequate digestion and ab-

sorption of fat. In the absence of the terminal ileum, increased bile acids will be delivered to the colon. Bile salts in the colon increase the water content of the feces by promoting increased secretion of water into the lumen of the colon, resulting in a watery diarrhea. Choice A is incorrect because current therapy in this patient involves folic acid supplements. VIP-secreting tumor (choice C) causes an excess watery secretion by intestinal glands, resulting in an overwhelming of the absorptive capacity of the colon and a watery diarrhea, but is not the result of ileal resection. Cholera toxin (choice D) stimulates cAMP production and a massive watery intestinal secretion and diarrhea, but there is no evidence of cholera in this case. Recurrent Crohn's disease (choice E) would produce diarrhea, but radiographic studies in this case reveal no evidence of recurrent disease.

6. C is correct. Total gastrectomy would result in the removal of gastric parietal cells, which are the source of intrinsic factor necessary for the absorption of vitamin B_{12}, which is required for red blood cell maturation. Vitamin D deficiency (choice A), vitamin K deficiency (choice B), vitamin A deficiency (choice D), and vitamin E deficiency (choice E) would be produced with fat malabsorption associated with bile acid or pancreatic lipase deficiency because all are fat-soluble vitamins.

7. B is correct. Intrinsic factor secreted by the parietal cells of the gastric mucosa combines with dietary vitamin B_{12} in the small intestine, and this intrinsic factor–vitamin B_{12} complex is carried to the terminal ileum where it is absorbed by active transport. Secretin (choice A) is a peptide hormone released by acidification of the small intestine and would not be affected by total gastrectomy. Pancreozymin (PZ) (choice C) is an old name for cholecystokinin (CCK) that was once known as CCK-PZ and is a hormone from the small intestine whose release is stimulated by dietary fat and protein. Antihemophilic factor A (choice D) would not be associated with anemia because this factor is linked to stopping blood loss. Pyridoxine (choice E) is a form of vitamin B_6 used in the treatment of vitamin B_6 deficiency, not vitamin B_{12} deficiency.

CHAPTER 6
ENDOCRINE PHYSIOLOGY

I. General Principles

A. Mechanism of Action of Hormones

1. **Hormones** are **chemical messengers secreted into the circulation by ductless glands.** Along with the nervous system, the endocrine system integrates organ systems via hormones secreted by endocrine tissues or glands.
2. **Water-soluble hormones** (ie, peptides and biogenic amines) attach to receptors on the plasma membrane of the target cell.
   a. Receptors for water-soluble hormones stimulate the production of intracellular second messengers (eg, cAMP, diacylglycerol, inositol 1,4,5-triphosphate, increased Ca^{2+}), which modify intracellular proteins (often enzymes) and bring about the hormone's biologic response.
   b. Water-soluble hormones circulate free (unbound) in the plasma and are continually available for degradation, thus accounting for their short plasma half-lives (generally 1–30 min).
3. **Lipid-soluble hormones** (ie, steroids and thyroid hormones) cross the plasma and nuclear membranes of their target cells readily and attach to receptors on the nuclear chromatin.
   a. The hormone receptor complex activates RNA polymerase, which transcribes a specific portion of the genome.
   b. Lipid-soluble hormones circulate bound to plasma proteins that serve as carriers. These carriers make the lipid-soluble hormones less available for degradation, thus accounting for their longer half-lives (usually hours for the steroid hormones and days for the thyroid hormones).
   c. Hormones can circulate free or bound to carrier proteins.
   d. Only **unbound hormones** can enter the target cell and initiate hormone action.
4. **Hormone receptors act as amplifiers of hormone action.** That is, one hormone-receptor complex can give rise to numerous copies of the second messenger molecule or of a newly synthesized protein.
   a. Under most normal physiologic conditions, the number of hormone receptors is not rate limiting for hormone action. Therefore, measurements of the plasma concentration of a hormone reflects the level of activity of the hormone.
   b. Generally, if excess hormone is present in the plasma, the number of receptors on that hormone's target cells decreases (**down-regulation**).

c. Hormones can have complementary and antagonist actions.
   (1) An example of a complementary action is the estradiol stimulation of progesterone receptors in the endometrium.
   (2) An example of antagonistic action would be insulin inhibition of gluconeogenesis and glucagon stimulation of gluconeogenesis.

B. Neuroendocrine Relationships

1. **The hypothalamus and pituitary gland provide central control of multiple endocrine organs.**
   a. The hypothalamic-hypophyseal portal veins provide a link by which the central nervous system modifies the rate at which specific, hypothalamic hormones are secreted into the portal vessels of the anterior pituitary.
2. Hypothalamic releasing hormones—**thyrotropin-releasing hormone** (**TRH**), **corticotropin-releasing hormone** (**CRH**), **growth hormone–releasing hormone** (**GHRH**), **somatostatin,** and **prolactin-inhibiting factor** (**PIF**)—are synthesized in neuronal cell bodies in the ventromedial, arcuate, and paraventricular nuclei (Table 6–1). **Gonadotropin-releasing hormone** (**GnRH**) is synthesized in the preoptic nucleus.
3. The nerve endings converge in the median eminence, and the hormones are then secreted into the hypophyseal-portal system and transported to the anterior pituitary.
4. Hypothalamic hormones bind to receptors on cells of the anterior pituitary and modify the secretion of **thyroid-stimulating hormone** (**TSH,** thyrotropin), **adrenocorticotrophic hormone** (**ACTH,** corticotropin), **luteinizing hormone** (**LH**), **follicle-stimulating hormone** (**FSH**), **growth hormone** (**GH**), and **prolactin.**
5. Most hypothalamic hormones promote the secretion of their respective pituitary hormone. Exceptions are **somatostatin,** which inhibits GH secretion, and **PIF,** which inhibits prolactin secretion.
6. **Endocrine regulation occurs through feedback control.** The **rate of secretion** of tropic hormones (eg, TSH, ACTH, LH, FSH) **is inversely proportional to the plasma concentration** of the hormone(s) secreted by their respective target glands.
7. Hormonal release is mainly pulsatile in the hypothalamic–anterior pituitary system.
8. Pulsatile release of GnRH prevents down-regulation of its receptors on the gonadotrophs of the anterior pituitary. Thus, a constant infusion of GnRH will decrease the release of both LH and FSH and is used to treat **precocious puberty.**
9. When removed from the influence of all hypothalamic target hormones, the anterior pituitary decreases its secretion of all hormones except **prolactin.**
10. The secretion of prolactin increases because a chronic source of inhibition (ie, **PIF**) has been removed.
11. Prolactin levels increase during pregnancy in response to the elevated concentrations of estrogen and progesterone. However, the lactogenic effect of prolactin appears to be inhibited by the elevated estrogen concentration during pregnancy.
12. Upon delivery, estrogen levels fall, thereby allowing the lactogenic effect of prolactin, and lactogenesis occurs.

Table 6–1. Hypothalamic hormones.

Hypothalamic Hormone	Purified	Synthesized	Stimulates	Inhibits	Location of Cell Bodies
Vasopressin	Yes	Nonapeptide	ACTH	0	SO, PVN
Corticotropin-releasing hormone (CRH)	Yes	41AA peptide	ACTH	0	PVN
LH-releasing hormone (LHRH)	Yes	Decapeptide	LH, FSH	0	POA, MBH
FSH-releasing factor	Yes	No	FSH	0	PVN/POA?
Thyrotropin-releasing hormone (TRH)	Yes	Tripeptide	TSH, PRL	0	PVN
Growth hormone-releasing hormone (GHRH)	Yes	44AA peptide	GH	0	ARC
Growth hormone secretagogue(s)	No	Only a synthetic hexapeptide is available	GH	0	?
Somatostatin	Yes	Tetradeca-peptide	0	GH, PRL, TSH, gastrin, glucagon, insulin	APR
Prolactin (PRL)-inhibiting factors (PIFs), dopamine,	Yes	Yes	0	PRL	ARC
peptidergic PIF	No	No	0	PRL	?
PRI,-releasing factors (PRFs)					
Prolactin-releasing peptide	Yes	20–31AA	PRL	0	DMH, SN
Oxytocin	Yes	Nonapeptide	PRL	0	SO, PVN
TRH, VIP, PHI, angiotensin II, neurotensin, substance P	Yes	—	PRI	0	PVN, SO
MSH-releasing factor (MRF)	Yes	No	MSH	0	
MSH-inhibiting factor (MIF)	Yes	No	0	MSH	
Pituitary adenylate cyclase-activating peptide (PACAP)	Yes	27 and 38AA	ACTH, GH	0	SO, PVN

APR, anterior periventricular region; ARC, arcuate nucleus; DMH, dorsomedial hypothalamus; MBH, medial basal hypothalamus; MSH, melanocyte-stimulating hormone; PHI, peptide histidine isoleucine; POA, preoptic area; PVN, paraventricular nucleus; SN, solitary nucleus; SO, supraoptic nucleus; VIP, vasoactive intestinal peptide; O, unknown; ?, not found yet; —, not synthesized yet.

13. After parturition, prolactin is stimulated primarily by nursing.
14. Regular nursing activity can maintain prolactin levels high enough to decrease GnRH secretion and inhibit ovulation.

HYPERPROLACTINEMIA

- *A **prolactinoma** is the most common pituitary tumor.*
- *Prolactinomas produce secondary **amenorrhea** and **galactorrhea** (persistent milk discharge in the absence of parturition) in women and **impotence** in men.*
- ***Bromocriptine,** a dopamine analogue, is most commonly used to inhibit prolactin release.*
- ***Primary hypothyroidism** (reduced secretion of thyroid hormone) also causes hyperprolactinemia due to increased **TRH,** a potent prolactin stimulator.*

C. Posterior Pituitary Hormones

1. **Unlike the anterior pituitary, the posterior pituitary is actually part of the brain.**
2. **Antidiuretic hormone** (**ADH**) or **Arginine vasopressin** (**AVP**) is synthesized in the hypothalamus, but it is stored and released from the posterior pituitary.
3. ADH increases water permeability of the renal collecting duct by placing water channels in the membrane.
   a. Water is **reabsorbed passively,** drawn across the membranes by the higher osmolarity of the interstitium.
   b. Urea, a lipid-soluble solute, can pass with the water, but electrolytes cannot.
4. ADH **secretion is controlled** primarily **by hypovolemia and plasma osmolality.**
   a. Decreased blood volume causes venous and arterial stretch receptors to send fewer signals to the central nervous system, decreasing chronic inhibition of ADH secretion. This mechanism is especially important for restoring extracellular fluid (ECF) volume following a hemorrhage.
   b. An increase of only 1% in the osmolality of the ECF bathing the hypothalamic osmoreceptors will evoke an increased rate of ADH secretion. In this manner, ECF osmolality is kept very close to 300 mOsm/L.
   c. ADH secretion is inhibited by ethanol ingestion and weightlessness.
5. **Oxytocin originates** primarily **in the paraventricular nuclei** of the hypothalamus and causes milk flow (ie, letdown) from the breast.
6. Oxytocin release is stimulated by suckling at the breast, sexual activity, or emotional factors (eg, hearing the infant cry).
7. Oxytocin causes contraction of the myoepithelial cells of the mammary gland and uterine contractions at term.

DIABETES INSIPIDUS

- *Excretion of large quantities of dilute urine with increased blood osmolality is diagnostic for diabetes insipidus (DI).*
- ***Central DI** is associated low ADH release.*
- ***Nephrogenic DI** is due to the inability of the kidneys to respond to ADH.*
- *Syndrome of inappropriate ADH secretion (**SIADH**) causes increased ADH secretion and excessive water reabsorption in the collecting ducts.*

II. Adrenal Cortex

A. Adrenal Hormones Secreted by the Adrenal Cortex

1. **Mineralocorticoids,** such as aldosterone, are produced in the zona glomerulosa, and primarily regulate salt balance.
2. **Glucocorticoids,** are synthesized from cholesterol in the zona fasciculata and reticularis, and have a metabolic function.
   a. Cortisol is the primary glucocorticoid hormone in humans.
3. **Sex steroids,** such as dehydroepiandrosterone (**DHEA**), are produced primarily in the zona reticularis, and have a reproductive function.

B. Synthesis of Adrenal Hormones from Cholesterol

1. The rate-limiting step is conversion of cholesterol to **pregnenolone.**
2. **Desmolase,** a mitochondrial cytochrome P-450 side chain cleavage enzyme, is responsible for the conversion.
3. **ACTH** stimulates desmolase activity in adrenocortical cells.
4. In the zona glomerulosa, **angiotensin II** also stimulates desmolase activity.

C. Aldosterone Production (Figure 6–1)

1. The glomerulosa cells synthesize aldosterone from cholesterol via **progesterone.**
2. In the smooth endoplasmic reticulum, **21-β-hydroxylase** converts progesterone to **11-deoxycorticosterone** (**DOC**), which has mineralocorticoid activity.
3. **11-β-hydroxylase** acts on DOC in mitochondria to form **corticosterone,** which has weak glucocorticoid activity and mineralocorticoid activity.
4. Two mitochondrial enzymes active only in glomerulosa cells, 18-hydroxylase and 18-hydroxydehydrogenase, convert corticosterone to **aldosterone.**

D. Cortisol Synthesis (see Figure 6–1)

1. **17-α-Hydroxylase,** a smooth endoplasmic reticulum enzyme, acts on pregnenolone to form 17-hydroxypregnenolone.
2. **17-Hydroxyprogesterone** is hydroxylated at C_{21} to form 11-deoxycortisol.
3. **11-Deoxycortisol** is subsequently hydroxylated at C_{11} to form cortisol, the most potent natural glucocorticoid in humans.

E. Adrenal Androgen Synthesis

1. 17-Hydroxypregnenolone is converted to **DHEA** by a smooth endoplasmic reticulum enzyme, 17,20 lyase.
2. Although DHEA is produced in great quantity, it is a relatively weak adrenal androgen. It serves mainly as a precursor for δ^4 androstene-3, 17-dione, which is a more potent androgen.
3. **Androstenedione** can be reduced at C_{17} to form testosterone.
4. From ages 5 to 13 years, corresponding with growth of the reticularis layer of the cortex, the production of adrenal androgens increases (**adrenarche**). Adrenal androgens play a primary role in the development of pubic and axillary hair in the female.
5. A number of congenital enzyme deficiencies in the pathways of adrenocortical hormone synthesis may occur, known as **adrenogenital syndrome.**

ADRENOGENITAL SYNDROME

- ***21-Hydroxylase deficiency is the most common enzymatic disorder that*** *accounts for this syndrome.*
- *ACTH secretion is increased because of the low cortisol production.*

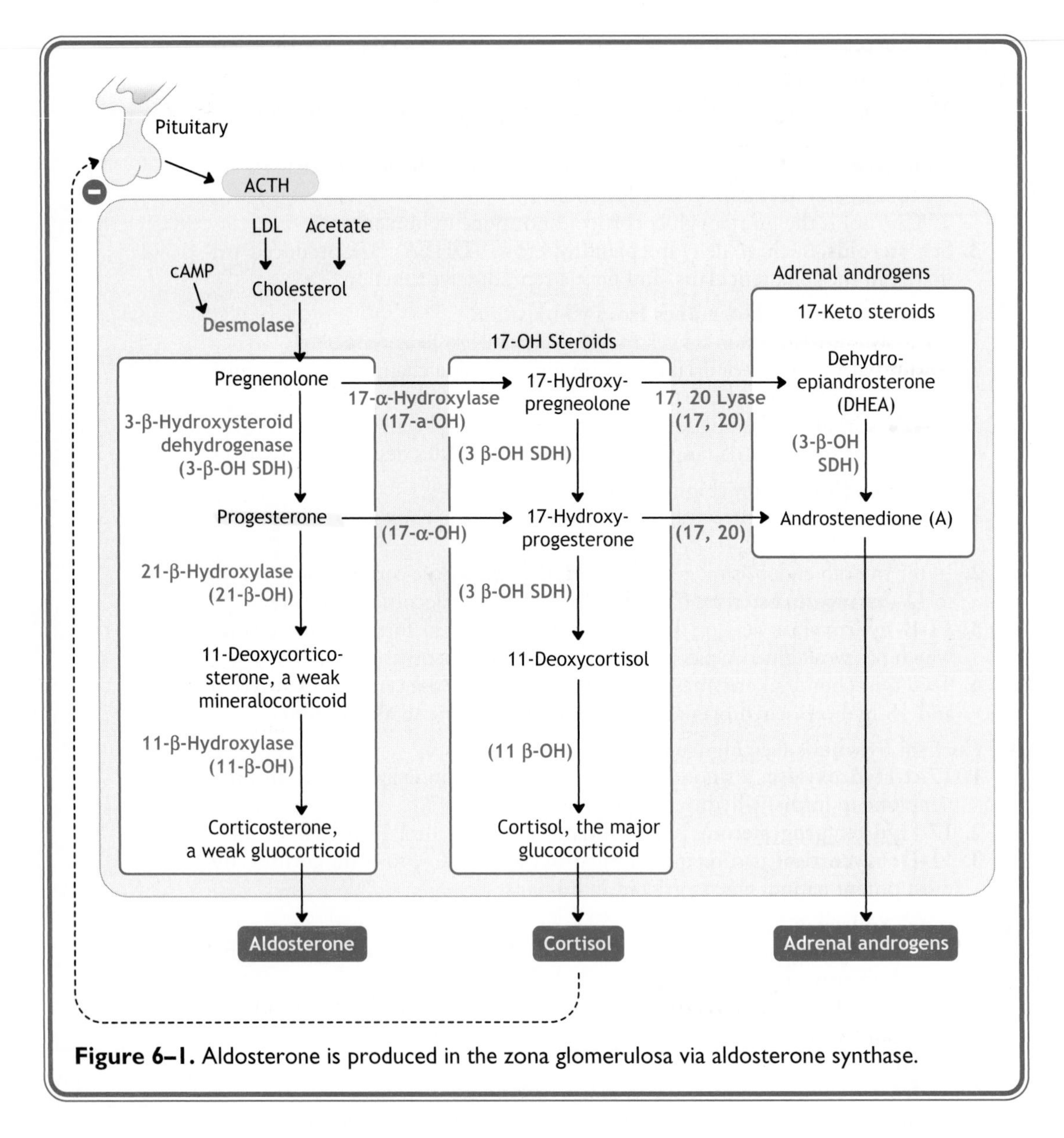

Figure 6–1. Aldosterone is produced in the zona glomerulosa via aldosterone synthase.

- *Adrenal androgens are produced in great excess, causing virilization.*
- *The clinical consequences can be dramatic. In females, ambiguous genitalia can lead to incorrect gender assignment at birth.*

F. Control of Adrenal Cortical Secretions

1. The **primary action of ACTH** is stimulation of desmolase for the **conversion of cholesterol to pregnenolone.**
2. Cortisol inhibits ACTH secretion at the pituitary and hypothalamic levels in a negative feedback manner whereas stress acts to stimulate the axis.

3. **CRH** from the hypothalamus stimulates ACTH secretion that in turn stimulates cortisol release from the adrenal cortex.
   a. Approximately 70% of daily cortisol release occurs between 12 AM and 8 AM. The peak release occurs between 6 AM and 8 AM.
   b. The low point in CRH release occurs during the evening.
4. Stress stimulates ACTH release. AVP is a potent ACTH secretagogue and may play a physiological role in ACTH secretion during stress.
5. ACTH also increases secretion of the adrenal androgens.
6. ACTH does not directly regulate aldosterone secretion but acts as a facilitator for primary regulators of aldosterone secretion, such as the renin-angiotensin system.

G. **Aldosterone Secretion**

1. **Aldosterone production is regulated by the renin-angiotensin system.**
   a. The **juxtaglomerular cells,** located in the walls of the afferent glomerular arteriole, respond as baroreceptors and secrete renin in response to changes in perfusion pressure.
   b. The **macula densa,** a specialized area near the juxtaglomerular cells, monitors tubular composition and mediates renin release.
   c. Factors that decrease fluid volume (eg, dehydration or blood loss) or decreased Na^+ concentration stimulate renin release.
   d. Renal sympathetic nerves that innervate the juxtaglomerular cells mediate certain effects on renin release, including central nervous system effects, such as hypovolemia, and postural effects. This mediation is independent of the renal baroreceptor and salt effects.
   e. Renin release also appears to be mediated by local prostaglandin production.
   f. The enzyme renin acts on the substrate angiotensinogen (an α_2-globulin of hepatic origin) to produce the decapeptide angiotensin I.
   g. **Angiotensin-converting enzyme** (**ACE**), found primarily in the lungs, removes two carboxy terminal amino acids from **angiotensin I,** producing **angiotensin II.**
      (1) Angiotensin II directly stimulates aldosterone production.
      (2) Angiotensin II also increases arteriolar vasoconstriction.
   h. In humans, angiotensin II can be further cleaved to **angiotensin III.**
   i. Aldosterone acts at distal tubule and collecting duct cells of the kidney to stimulate the **active reabsorption of Na^+ and the tubular secretions of K^+ and H^+.**
   j. Aldosterone exerts indirect negative feedback on the renin-angiotensin system by increasing the circulating volume and lowering plasma K^+ concentration.
   k. **Atrial natriuretic peptide,** which is synthesized and released by atrial myocytes in response to increased vascular volume, can decrease aldosterone secretion.

PRIMARY AND SECONDARY ALDOSTERONISM

- *Adenomas of the glomerulosa cells can result in **primary aldosteronism.** Manifestations include hypertension, hypokalemia, hypernatremia, and alkalosis.*
- *In patients with renal artery stenosis, edema, and **secondary aldosteronism,** similar manifestations are noted, along with increased renin and angiotensin II levels.*

H. Metabolic Actions of Glucocorticoids

1. **Glucocorticoids** promote the mobilization of energy stores, specifically **amino acids** from protein, and **free fatty acids** and **glycerol** from the triglycerides of adipose tissue.
2. They inhibit glucose uptake in most tissues (ie, muscle, lymphoid, and connective tissue), thereby sparing plasma glucose for brain metabolism.
3. While inhibiting amino acid uptake and protein synthesis in most tissues, glucocorticoids promote protein breakdown.
4. The increased delivery of amino acids to the liver and the increased activity of liver gluconeogenic enzymes and glucose-6-phosphatase allows for increased generation of glucose from protein stores (**gluconeogenesis**).
5. When glucocorticoids are present in excess, liver **glucose-6-phosphate** production is increased so much that glycogen formation is also increased.
6. Cortisol enhances the capacity of glucagon and catecholamines to act.
   a. Glucagon and epinephrine promote **glycogenolysis** and lipolysis, but cortisol must be present for glucagon and epinephrine to exert their full glycogenolytic and lipolytic effects.
   b. The catecholamines **epinephrine** and **norepinephrine** promote vasoconstriction and bronchodilation, but cortisol must be present for these effects to be manifested fully.
7. Cortisol secretion and the resistance to the physiologic impact of stress are linked.
   a. The capacity to withstand stress depends on adequate glucocorticoid secretion.
   b. Stress is a potent stimulator of CRH, ACTH, and cortisol.

CORTISOL EXCESS (CUSHING SYNDROME)

- *Cushing syndrome is a manifestation of hypercortisolism most commonly due to long-term glucocorticoid therapy.*
- *Ninety percent of patients exhibit weight gain, abnormal fat distribution in the face (moon face), and upper back (buffalo hump) and truncal obesity.*
- *Many patients exhibit diastolic hypertension and glucose intolerance.*
- *Decreased collagen synthesis and increased collagen breakdown cause thinning of skin, resulting in stretch marks (**purple striae**).*
- *Increased bone resorption and decreased bone formation result in **osteoporosis.***
- *Increased **lymphopenia** (reduced number of lymphocytes in blood) results in decreased antibody production and poor wound healing.*
- *The best screening test is 24-hour urinary free cortisol, which measures excess unbound cortisol.*
- *Cushing syndrome is associated with decreased ACTH, whereas Cushing disease (a pituitary disorder) is associated with increased ACTH levels.*

ADRENAL INSUFFICIENCY (ADDISON DISEASE)

- *Autoimmune destruction of the adrenal glands is the most common cause of Addison disease.*
- *Chief clinical findings include hypotension, muscle weakness, anorexia, weight loss, and diffuse hyperpigmentation.*
- *The disorder is associated with elevated plasma ACTH, hyponatremia, hyperkalemia, fasting hypoglycemia, and eosinophilia.*
- *Treatment involves replacement of glucocorticoids.*

III. Adrenal Medulla

A. The adrenal medulla is essentially a **specialized sympathetic ganglion** that links the endocrine and sympathetic nervous systems.
   1. The adrenal medulla is innervated by cholinergic preganglionic fibers.
   2. The primary secretory product is epinephrine.
      a. Adrenal medullary chromaffin cells are the only ones capable of synthesizing epinephrine.
   3. Norepinephrine is the primary catecholamine of the sympathetic nervous system.
   4. Catecholamines bind to α and β adrenoreceptors on cell surface receptors and act through heterotrimeric G proteins.

B. The actions of epinephrine, norepinephrine, and dopamine at a particular tissue depend on the types of receptors present, their affinity for the catecholamine, and the catecholamine involved (Figure 6–2).

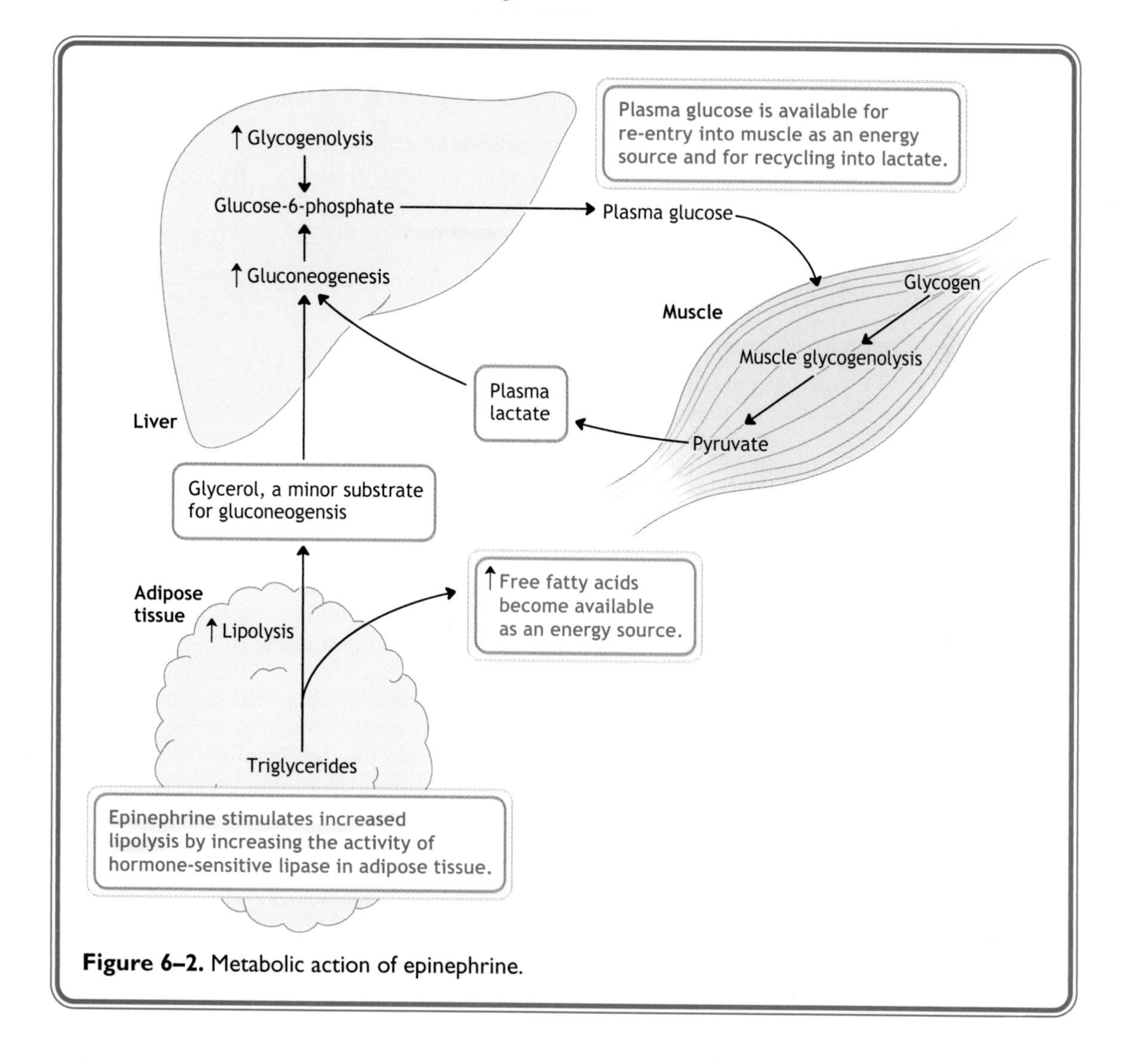

Figure 6–2. Metabolic action of epinephrine.

1. The three classes of adrenergic receptors are associated with different activities.
   a. **α-Adrenergic receptors** mediate vasoconstriction.
   b. **β_1-Adrenergic receptors** mediate cardiac inotropic and chronotropic effects.
   c. **β_2-Adrenergic receptors** mediate bronchiolar smooth muscle relaxation.
2. A synthetic agonist that acts on β-receptors is isoproterenol.
3. Dopaminergic D_1-receptors mediate renal vasodilation.
4. Dopaminergic D_2-receptors are associated with nausea and prolactin-release inhibition.

PHEOCHROMOCYTOMA AND NEUROBLASTOMA

- *A **pheochromocytoma** is a relatively uncommon catecholamine-secreting tumor.*
  –Approximately two-thirds of patients exhibit a sustained hypertension, and the other one-third experience episodic hypertension (eg, upon standing or with stress).
  –If epinephrine is the primary secretory product, the heart rate usually will be increased.
  –If norepinephrine is the primary secretory product, the heart rate will be decreased reflexly in response to marked peripheral vasoconstriction.
  –Pheochromocytomas can be benign or malignant unilateral adenomas of the adrenal medulla.
- *A **neuroblastoma** is a malignant small cell tumor of neural crest origin commonly developing in the adrenal medulla of children.*
  –Chief clinical findings include palpable abdominal mass, diastolic hypertension, and elevated urinary catecholamines.
  –The prognosis depends on age; the cure rate is high in children under 1 year and low in older children.

IV. Endocrine Pancreas

A. The Islets of Langerhans are endocrine and paracrine tissues.

1. **α cells** produce the hormone **glucagon** and make up 20–25% of pancreatic islets.
2. **β cells** produce **insulin** and make up 65–75% of pancreatic islets.
3. **δ cells** produce **somatostatin** and make up 5% of pancreatic islets.
4. **F cells** secrete pancreatic polypeptide.

B. Insulin Production and Secretion

1. Insulin is **initially synthesized as a pre-prohormone.** A 23-amino-acid *N*-terminal signal peptide, the B, C, and A chains are translated in that order.
2. The signal peptide directs the forming molecule into the cisternae of the endoplasmic reticulum and is then removed.
3. Removal of the signal peptide leaves the proinsulin molecule, which provides the proper conformation for formation of the disulfide bridges.
4. In the Golgi, specific proteases cleave the proinsulin molecule into **insulin** and **C-peptide** and the insulin and C-peptide are packaged in equimolar amounts into secretory granules.
5. Although C peptide has no biologic action, it is secreted in a 1:1 molar ratio with insulin and serves as a useful marker for insulin secretion.
6. In response to Ca^{2+}, the granules move to the plasma membrane, where their contents are released by exocytosis.
7. The **primary stimulus for insulin release is blood glucose** that is metabolized by the β cell.

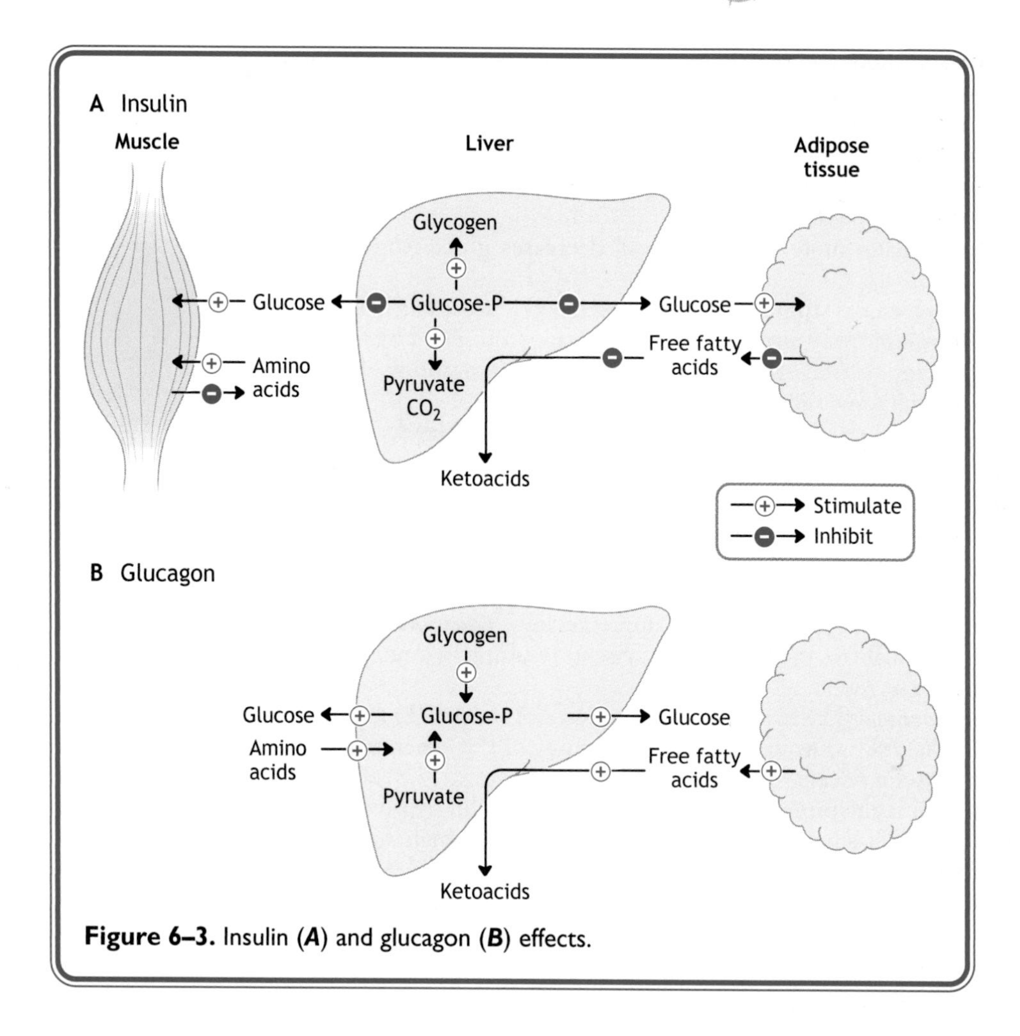

Figure 6–3. Insulin (***A***) and glucagon (***B***) effects.

C. Specific Actions of Insulin

1. Insulin **promotes fat deposition** and storage in adipose tissue (Figure 6–3A).
   a. Insulin **stimulates lipoprotein lipase activity,** causing breakdown of triglycerides from very low density lipids and chylomicrons to free fatty acids, which are taken into adipose tissue.
   b. Insulin **stimulates glucose uptake in adipose tissue.** Glycerol-P is necessary for triglyceride synthesis, and its production is dependent on glucose uptake.
   c. Insulin **inhibits** hormone-sensitive lipase, decreasing **lipolysis** and release of free fatty acids to the plasma.
   d. **Ketoacid formation is inhibited** by decreased fatty acid degradation.
2. **In muscle, insulin stimulates amino acid uptake and protein synthesis,** while decreasing proteolysis and release of amino acids in blood to inhibit gluconeogenesis.
3. In the liver, insulin **promotes glucose storage as glycogen, glycolysis and synthesis of fatty acids.**

a. Insulin **promotes glucokinase activity,** trapping glucose in cells as glucose 6-phosphate.
b. Insulin activates the glycogen synthase enzyme complex and inhibits phosphorylase activity **to promote glycogen formation.**
c. Insulin also **decreases glucose-6-phosphatase,** decreasing release of glucose from the liver.

4. Insulin stimulates protein synthesis and **decreases gluconeogenesis in the liver.**
5. Uptake of glucose is stimulated, decreasing blood glucose concentration.
6. **Phosphofructokinase and pyruvate kinase are stimulated** by insulin.
7. **In adipocytes,** insulin **stimulates lipogenesis** by promoting the uptake of glucose and its conversion to triglycerides for storage.
8. Insulin increases cellular K^+ uptake, decreasing blood K^+ levels.
9. Insulin also favors hepatic sequestration of cholesterol by activating hydroxymethylglutaryl CoA reductase, the rate-limiting enzyme in cholesterol synthesis.

D. Stimulants of Insulin Release (Table 6–2)

1. **Glucose is the major regulator** of insulin release.
   a. Normal plasma glucose levels are approximately 90 mg/dL.
   b. In type II diabetes the fast phase is absent, resulting in a more gradual rise in insulin secretion.
   c. Glucose enters β cells via glucose transporter 2 (GLUT 2) and is metabolized. The ATP generated causes the opening of Ca^{2+} channels and subsequent insulin release.
   d. Glucagon is the primary counter hormone to insulin by possessing opposite effects. Plasma glucagon is low when glucose is high, and high when glucose is low. Glucagon increases plasma glucose levels.
2. **Amino acids** stimulate insulin secretion. They also stimulate glucagon release, which counters the effect of insulin, keeping blood glucose levels constant.
3. **Gastrointestinal (GI) hormones,** such as gastric inhibitory peptide (**GIP**) and glucagonlike peptide-1, stimulate insulin secretion.
4. **Glucagon** increases insulin secretion.
5. **Vagal stimulation** via acetylcholine release during feeding and acetylcholine administration will increase insulin release.
6. **Theophylline** is a phosphodiesterase inhibitor that increases cAMP in β cells, which leads to increased insulin secretion.
7. **Sulfonylureas** (oral hypoglycemic drugs) lower blood glucose by stimulating insulin secretion. They are useful in treating type II diabetes (where insulin is present) but are not effective in treating type I diabetes.
8. **Salicylates** can inhibit cyclooxygenase and block the inhibition of insulin release exerted by prostaglandins.

E. Inhibitors of Insulin Release

1. **α_2-adrenergic stimulation (by norepinephrine)**, particularly during **exercise, inhibits insulin secretion.**
   a. In a stressful situation (eg, infection) patients with diabetes are at greater risk for development of hyperglycemia.
   b. Increased sympathetic activity (epinephrine and norepinephrine) causes inhibition of endogenous insulin secretion.

Table 6–2. Factors affecting insulin secretion.

Stimulation	Inhibition
Physiologic	
Glucose	
Amino acids	
Gastrointestinal peptide hormones (esp. GIP)	
Ketone bodies (esp. in starvation)	
Glucagon	Somatostatin
Parasympathetic stimulation	Sympathetic stimulation (splanchnic nerve)
β-Adrenergic stimulation	α-Adrenergic stimulation
Pharmacologic and experimental	
Cyclic AMP	α-Deoxyglucose
Theophylline	Mannoheptulose
Sulfonylureas	Diazoxide
Salicylates	Prostaglandins Diphenylhydantoin β cell poisons: alloxan, streptozotocin

c. Sympathetic stimulation causes a β_2-adrenergic effect, which stimulates insulin release.

2. **Somatostatin** inhibits both insulin and glucagon secretion.
3. **Diazoxide,** an antihypertensive drug, is a potent inhibitor of insulin secretion.
4. **Prostaglandins** can inhibit insulin secretion.
5. **Diphenylhydantoin** is an anticonvulsant drug that suppresses insulin release.

F. Insulin Receptors

1. An insulin receptor contains two **α** and two **β subunits.**
2. The β subunits have **tyrosine kinase activity.**
3. **Because** high levels of insulin down-regulates insulin receptors, **starvation increases** and **obesity decreases** the number of insulin receptors.

V. Glucagon

A. Glucagon Structure

1. Glucagon is a single-chain peptide hormone.

2. It contains 29 amino acids secreted by pancreatic islet α cells.
3. Pancreatic α cells secrete glucagon in response to ingested protein

B. Glucagon Actions

1. Glucagon is a potent glucogenic hormone that promotes the synthesis of glucose by the liver and **increases blood glucose levels.** (Figure 6–3B).
2. Glucagon **stimulates lipolysis,** especially when insulin levels are low which can lead to formation of ketone bodies.
3. Thus, **Glucagon is a ketogenic hormone; insulin is an antiketogenic hormone.**

C. Control of Glucagon Secretion

1. **Decreased blood glucose is the major stimulator** of glucagon secretion.
2. **Protein intake** increases glucagon secretion.
3. **Exercise** increases glucagon secretion.
   a. Exercise causes increased uptake of glucose by muscle, decreasing blood glucose levels.
   b. Diabetic patients must adjust their insulin dosage or carbohydrate intake when anticipating an increase in physical activity.
4. **Stress,** such as trauma or surgery, is a potent stimulator of glucagon secretion.
5. **Parasympathetic stimulation** (through acetylcholine release) causes increased glucagon secretion.
6. During **fasting,** rising glucagon levels promote production of ketone bodies and promote gluconeogenesis from muscle protein.
   a. Hypoglycemia is the most important stimulator of glucagon secretion.
   b. Hyperglycemia is the most important inhibitor.

D. Relationship of Insulin, Somatostatin, and Glucagon Secretion

1. Experimentally, somatostatin release from the pancreas can be stimulated by agents such as glucose, amino acids, cholinergic stimuli, and GI hormones.
2. After ingestion of a mixed meal in humans, plasma somatostatin levels increase modestly.
3. Somatostatin is capable of inhibiting secretion of insulin and glucagon and other hormones (Figure 6–4).
4. Glucagon stimulates release of both insulin and somatostatin.
5. Insulin inhibits glucagon release and may have some as yet unknown effect on somatostatin.

DIABETES MELLITUS

Diabetes has been described as "starvation in the midst of plenty." Plasma glucose is increased, but utilization of glucose by most tissues of the body is depressed.

- *The three "polys" of diabetes are as follows:*
  *–**Polyuria:** The filtered glucose load exceeds the tubular capacity of the kidney for reabsorption, creating an osmotic diuresis.*
  *–**Polydipsia:** Increased thirst is stimulated by the hyperosmolality of the plasma and the resulting hypovolemia.*
  *–**Polyphagia**: Deficient glucose utilization in hypothalamic ventromedial nuclei cells causes the patient to eat more.*
- *The response to a standard oral test dose of glucose, in which 75 g of glucose are ingested and blood samples for measurement of plasma glucose are obtained 2 hours later, is called the glucose tolerance*

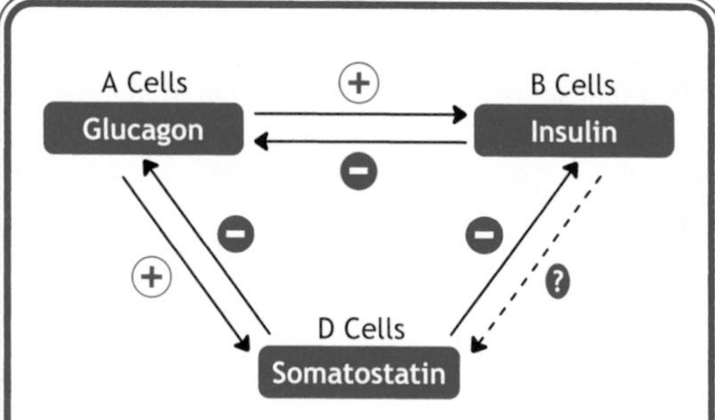

Figure 6–4. Interaction of glucagon, insulin, and somatostatin. Somatostatin is a powerful inhibitor of both glucagon and insulin. Glucagon stimulates insulin and somatostatin release. Insulin produces an inhibitory signal by unknown means to inhibit glucagon release and has unknown effects on somatostatin release.

test and is used in the clinical diagnosis of diabetes. In overt diabetics, fasting blood glucose values are elevated, and a persistent hyperglycemia is noted following ingestion of the test dosage of glucose.

There are two common types of spontaneous diabetes mellitus (DM):

- ***Type 2 DM,** formerly known as adult-onset or non-insulin-dependent diabetes, accounts for nearly 90% of all cases of diabetes.*
  –It occurs primarily in adults (after age 40) with a strong family history of the disease and is associated with increased insulin resistance (caused mainly by obesity). Type 2 DM is caused by abnormalities in both insulin secretion and insulin action.
  –Early type 2 DM is associated with normal β-cell morphology and insulin content.
  –Plasma insulin levels are often elevated, with a delayed but exaggerated and prolonged response to glucose.
  –Insulin resistance is caused by prereceptor defects that prevent insulin from binding to its receptor, receptor defects that cause decreased receptor number, or postreceptor defects that prevent the receptor from mediating insulin effects in the cell.
  –Individuals with type 2 DM tend to overeat, and the increased stimulation of insulin secretion results in decreased insulin sensitivity by target tissues. A compensatory response is a further increase in insulin secretion.
  –Type 2 DM occurs when the pancreatic reserve is exceeded.
  –The majority of patients are obese, and their glucose tolerance can be restored to normal with a controlled diet and exercise. Weight loss decreases insulin resistance.
  Type 2 DM appears to have a strong genetic component. Concordance rates in twin studies are close to 100%.
- ***Type 1 DM,** formerly known as juvenile-onset or insulin-dependent diabetes, occurs primarily in juveniles, is most often due to pancreatic islet β-cell destruction by an autoimmune process, and is associated with ketoacidosis. It is caused by an insulin deficiency and is not associated with obesity.*
  –Plasma insulin is low with abnormal β-cells at the time of diagnosis.
  –Plasma glucagon is increased, despite an elevated level of glucose.
  –Type 1 DM is now believed to be an autoimmune disorder, and patients treated early with immunosuppressive drugs, such as cyclosporin, show marked improvement.

–Most children with type 1 DM have antibodies to islet cells, and many have antibodies to glutamic acid decarboxylase thought to be important in initiating the destruction of β cells.
–Insulin reactions, which cause incoordination and slurred speech, are common in individuals with type 1 DM.
–Ketoacids increase and raise the hydrogen ion concentration. If blood pH falls to low levels, coma can ensue and death from ketoacidosis can occur.
–Type 1 DM concordance rates are only 33% in twin studies. Thus, type 1 DM does not exhibit a strong genetic component.

- ***Hypoglycemia** the opposite to diabetes mellitus has many causes.*
  –Most frequent cause is Type 1 DM patient who skips a meal or falls to adjust his insulin dosage when exercising.
  –β-adrenergic agonist drugs commonly cause hypoglycemia in diabetic patients taking insulin.
  –Alcoholic patients are not at risk for hypoglycemia since ethanol suppresses gluconeogenesis.
  –An insulinoma is a rare islet cell tumor (usually benign) that releases high concentrations of insulin into the blood stream.
  *–**Postprandial hypoglycemia** may occur in some people due to too much insulin being released too late after a meal.*

VI. Human Growth Hormone

A. Anterior Pituitary Hormone

1. **GH** a polypeptide hormone secreted by anterior pituitary somatotrophs is the principal endocrine regulator of somatic size.
2. GH, prolactin, and chorionic somatomammotropin constitute a family of hormones having considerable sequence homology. All have growth-promoting and lactogenic activity.

B. Actions of Growth Hormone

1. The most dramatic effect of GH is **stimulation of postnatal linear growth.** The principal mediator of this growth-promoting action is (**insulinlike growth factor 1,** IGF-1), whose production by the liver, cartilage, and other tissues is stimulated by GH.
2. **IGF-1** and **somatomedin A** (**IGF-2**) compounds have structures similar to **proinsulin** and actions similar to **insulin.**
   a. IGF-2 has actions similar to IGF-1 but is less dependent on GH.
3. GH increases amino acid uptake and **increases protein synthesis** in muscle.
4. GH has acute "anti-insulin" metabolic effects by **decreasing glucose** uptake in muscle and adipose tissue, and **increasing gluconeogenesis.**
5. **GH stimulates lipolysis.**

C. Control of GH Secretion (Figure 6–5)

1. Insulin-induced hypoglycemia is a clinical procedure for evaluating the pituitary reserve of GH.
2. **Hyperglycemia** suppresses GH secretion.
3. Hyperglycemia stimulates insulin release and the resulting **hypoglycemia** stimulates GH secretion.
4. **Amino acids** stimulate GH secretion; **arginine** is the most potent.
5. There is negative feedback regulation of GH. With decreased plasma levels of IGF-1, GH secretion increases and high IGF-1 levels decrease further release.
6. **Somatotrophs secrete** GH in a pulsatile fashion with the largest spikes occurring in stage IV sleep.
7. Strenuous **exercise,** such as running and bicycling, increases GH secretion.

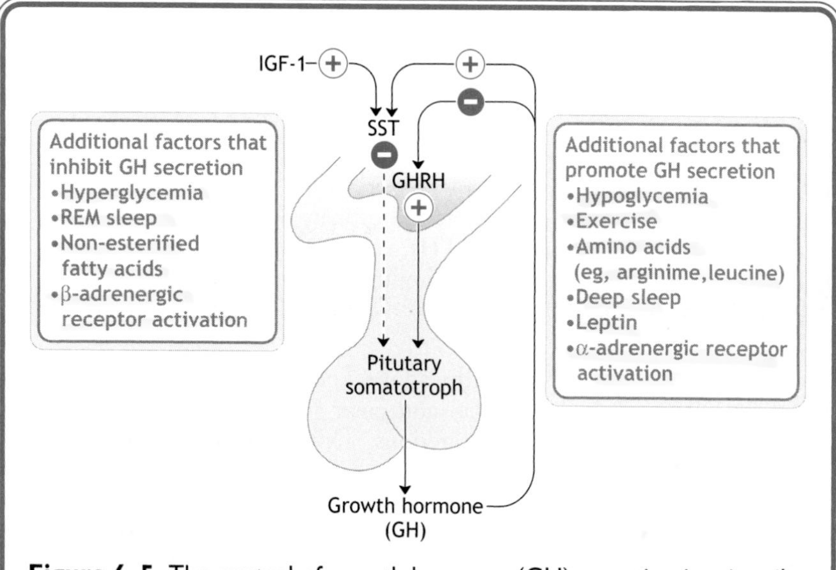

Figure 6–5. The control of growth hormone (GH) secretion is primarily via regulation of somatostatin (SST) and growth hormone–releasing hormone (GHRH). Insulinlike growth factor (IGF-1) and GH stimulate SST, which inhibits GH (dashed line). GH also negatively feeds back to inhibit GHRH, which normally stimulates GH secretion.

8. **Stress** increases GH secretion.
9. **Glucocorticoids, sex steroids, thyroid hormones,** and **insulin** also promote growth.
10. **Somatostatin** inhibits GH secretion.
11. **High progesterone** levels observed in late pregnancy may inhibit GH secretion.

D. Insulin, IGF-1, and IGF-2 Binding to Specific Receptors
1. Binding of IGF-1 to the insulin receptor can cause insulin effects, and binding of insulin to the IGF-1 receptor can cause IGF-1 effects.
2. The receptor for IGF-2 is a single-chain polypeptide that does not exhibit tyrosine kinase activity.

E. Regulation of IGF-1
1. GH provides normal physiologic regulation of IGF-1.
2. Fasting and insulin deficiency decrease IGF-1 production and release, despite elevated GH levels.
3. In late pregnancy, GH secretion is low, but the IGF-1 level is normal.
4. IGF-2 is thought to play a role in fetal development.

DWARFISM, GIGANTISM, AND ACROMEGALY

- ***Pituitary Dwarfism*** *caused by GH deficiency early in life.*
  *–**Laron dwarfism** is characterized by high plasma levels of GH and low plasma levels of IGF-1, because of a deficiency of GH receptors.*
  *–In **pygmies,** GH receptors are present, thus some other defect is involved.*

- *An increase in GH in childhood (which increases IGF-1) causes **gigantism.***
- *Increased GH after puberty causes **acromegaly**—an increase in periosteal bone growth of the chin, hands, and feet and enlargement of organs. Acromegaly is also associated with diastolic hypertension and diabetes mellitus.*

VII. Hormonal Calcium Regulation

A. Hormones in Calcium Regulation

1. The gut, kidneys, and bone regulate calcium and phosphate balance.
2. Three hormones; **parathyroid hormone (PTH)**, **calcitonin,** and **calcitriol** (**1,25$(OH)_2D_3$**)—the hormonally active metabolite of vitamin D_3 are involved.

B. Calcium Balance

1. Calcium intake averages 1000 mg/d in healthy adults, and about 350 mg are absorbed from the GI tract (small intestine) and enter the ECF each day. Roughly 150 mg are returned in gastric secretions. Thus, **net absorption equals about 200 mg/d.**
2. Normally, 98% of the calcium filtered by the kidney is reabsorbed. Urinary loss equals about 200 mg/d. Thus, intestinal absorption is balanced by urinary excretion.
3. Nearly 99% of total body calcium is found in bone.
4. A rapidly exchangeable pool of skeletal calcium is in equilibrium with calcium in the ECF.

C. Intestinal Absorption of Calcium

1. **Vitamin D by acting on the small intestine promotes bone mineralization.**
   a. In vitamin D deficiency, calcium absorption from the GI tract is decreased.
   b. If vitamin D is added to the diet, active calcium absorption occurs.

D. Plasma Calcium

1. Plasma calcium exists in three forms:
   a. Approximately 40% is bound to plasma protein, (primarily albumin).
   b. 10% of the total is bound to anions such as PO_4^{3-}, citrate, and isocitrate.
   c. Ionized or free Ca^{2+} makes up about 50% of the total and is the active plasma fraction.
2. The equilibrium between ionized and protein-bound calcium depends on blood pH. Acidosis decreases binding and increases ionized Ca^{2+}, whereas alkalosis increases binding and decreases ionized Ca^{2+}.

HYPOCALCEMIA

- *Low extracellular free Ca^{2+} levels can result in **hypocalcemic tetany.***
- *Extensive **spasms of the skeletal muscles** can occur.*
- ***Laryngospasm** can become so severe that the airway passage is obstructed and fatal asphyxia produced.*

E. Bone Cells

1. Bone remodeling depends on the coupled activities of osteoblasts and osteoclasts.
2. **Osteoblasts** are found on the surfaces of bone and are primarily responsible for bone formation.

3. Osteoblasts become surrounded by bone matrix. Bone cells surrounded by calcified matrix are called osteocytes and they send processes into the canaliculi, which spread throughout the bone.
4. **Osteoclasts** are large multinucleated cells responsible for bone resorption.
   a. **Osteoprotegerin** is a protein that appears to be a major stimulant of differentiation and activity of osteoclasts.
5. The osteoid of bone is composed of collagen and mucopolysaccharides.
6. In addition to Ca^{2+}, recently mineralized or partially mineralized bone contains ions such as PO_4^{3-}, Mg^{2+}, Na^{2+}, Cl^-, and K^+, which can be transferred to the ECF.

F. Parathyroid Hormone

1. PTH is a polypeptide produced in the parathyroid glands that is regulated by plasma Ca^{2+}.
2. Secretion is stimulated by a low plasma Ca^{2+} level and suppressed by a high plasma Ca^{2+} level.
3. The primary skeletal action of PTH is **stimulation of osteoclastic bone resorption.**
   a. In addition to stimulation of osteoclast activity, PTH also increases osteoclast numbers.
   b. Resorption of bone by osteoclasts is dependent on an increased H^+ concentration in the resorption zone and an increased release of lysosomal enzymes to promote breakdown of the bone matrix.
4. In the kidney, **PTH promotes** PO_4^{3-} loss, kidney tubular reabsorption of Ca^{2+}, and **1-hydroxylation of 25-hydroxyvitamin D** (**calcitriol**).
5. PTH stimulates **increased intestinal Ca^{2+} absorption.**

G. Calcitriol

1. **7-Dehydrocholesterol** is converted to cholecalciferol (vitamin D_3) in the skin by sunlight (Figure 6–6).
2. **Vitamin D_3** is transported by a sterol binding protein to the liver where it is hydroxylated to form 25-(OH)D3, the primary circulating form of vitamin D_3.
3. **25-(OH)D_3** is transported to the kidney where hydroxylation occurs to form calcitriol, the active metabolite of vitamin D_3.
4. **Deficiency** of calcitriol in children causes **rickets** (uncalcified osteoid) and vitamin D deficiency in adults produces **osteomalacia.**
5. Factors that increase formation of vitamin D_3 include
   a. PTH
   b. Low serum Ca^{2+}
   c. Phosphate ingestion
   d. Prolactin, GH, and insulin (though not as primary regulators)
6. **Calcitriol functions** include
   a. Increased Ca^{2+} and PO_4^{3-} absorption by the small intestine
   b. Increased Ca^{2+} reabsorption in the kidney distal convoluted tubule.
   c. Stimulation of bone resorption with high concentrations
7. Calcitriol mediates the PTH effect of increased intestinal Ca^{2+} absorption.

H. Calcitonin

1. Calcitonin is a polypeptide synthesized by parafollicular C cells of the thyroid gland.

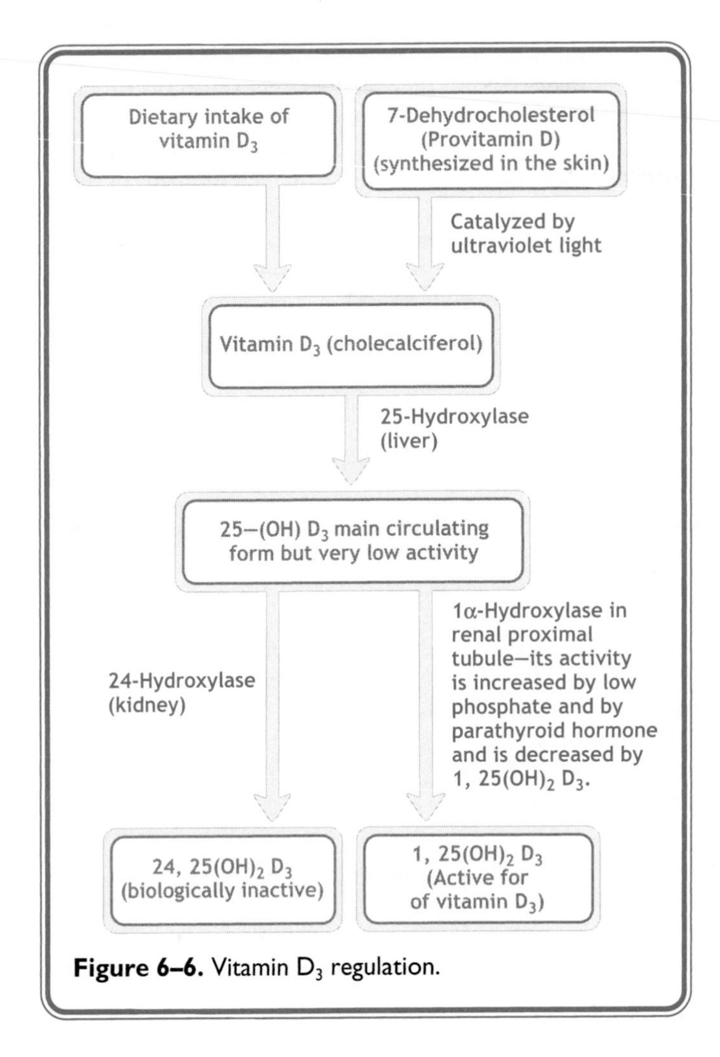

Figure 6–6. Vitamin D_3 regulation.

2. **Calcitonin inhibits osteoclasts** but it does not appear to be physiologically important in humans.
3. The nervous system releases **calcitonin gene–related peptide** (**CGRP**), a potent vasodilator substance.

DISORDERS AFFECTING BONE FORMATION

- ***Paget disease*** *is characterized by increased osteoclastic activity. Calcitriol and bisphosphonates have been used clinically to decrease the high bone turnover associated with this condition.*
- ***Osteoporosis*** *is a decreased mass of cortical and trabecular bone caused by a decrease in bone matrix.*
  –Since sex steroid hormones promote bone deposition, the major risk factor for osteoporosis is decreased estrogen levels in aging women.
- ***Agents for preventing or inhibiting the development of osteoporosis include antiresorptive drugs (bisphosphonates, calcitonin, estrogen) and drugs that stimulate bone formation (calcitriol and sodium fluoride).***

- ***Hypoparathyroidism*** *is due to or decreased function of the parathyroid glands, resulting in low PTH, which is characterized by hypocalcemia, hyperphosphatemia, and decreased PO_4^{3-} excretion.*
- ***Primary hyperparathyroidism*** *is usually due to a single adenoma, causing increased secretion of PTH and resulting in*
  *–Increased total plasma calcium (**hypercalcemia**)*
  –Increased ionized Ca^{2+} in the plasma
  –Decreased plasma PO_4^{3-} (hypophosphatemia)
  –Increased urinary PO_4^{3-} excretion (phosphaturia)
  –Increased urinary Ca^{2+} when the plasma Ca^{2+} increases beyond the concentration the kidney can reabsorb

VIII. Thyroid Hormones

A. Formation of the Thyroid Hormones

1. Thyroid hormones are made by iodinating tyrosine residues on thyroglobulin.
2. Iodide is oxidized to iodine by **thyroperoxidase.**
3. Iodine is incorporated into tyrosine residues to form monoiodotyrosine (**MIT**) or diiodotyrosine (**DIT**).
4. Two DITs are coupled to form **triiodothyronine (T_3)**.
5. One MIT and two DITs are coupled to form T_3, or **reverse T_3 (rT_3)**.
6. In the thyroid gland, most of the iodine is found in MIT and DIT and is stored as part of thyroglobulin molecules in thyroid follicles.
7. Thyroxine (T_4) is released in larger amounts than T_3, the primary active hormone. rT_3 is released but is inactive.

B. Iodine Metabolism

1. Because iodide is not in great abundance, the body has sensitive mechanisms of trapping and transporting the ion.
2. Only the thyroid gland significantly incorporates iodine into protein.
3. **Iodide** is **transported into the thyroid** at the basal membrane of follicular cells by an active process that may involve Na/K^+-ATPase in the trapping mechanism.
4. Iodide is **oxidized to iodine** in the follicular lumen.
   a. Iodine is incorporated into tyrosine residues (organification) to form **MIT** and **DIT.**
   b. Coupling of MIT and DIT to form **T_4** and **T_3** occurs.
5. T_4 and T_3 are stored as part of thyroglobulin residues in thyroid gland.
6. **TSH** stimulation causes the residues to be taken into the follicular cells by endocytosis.
7. In the follicular cell, thyroglobulin residues are acted on by lysosomal and protease enzymes to form T_4 and T_3, that are released into the blood for binding to thyroid binding globulin (TBG) and other proteins.

C. Thyroid Hormone Transport

1. TBG is a glycoprotein produced by the liver that binds approximately 70% of the T_4 and approximately 80% of the T_3 in plasma.
2. Albumin has a much lower affinity for binding the thyroid hormones than does TBG, but the high concentration of this protein results in the binding of approximately 20% of T_4 and approximately 11% of T_3.
3. Alterations of thyroid hormone–binding proteins can alter total thyroid hormone levels.

a. During pregnancy, increased estrogen causes increased TBG production. Pregnant women do not become hyperthyroid, however, because free thyroid hormone levels remain relatively constant.
b. Numerous drugs (eg, salicylates) compete with T_4 and T_3 for binding sites on TBG, producing low total thyroid hormone levels, but free hormone levels again remain relatively normal.
c. Peripheral tissues deiodinate T_4 to produce T_3.

D. Feedback Regulation of Thyroid Function

1. Although the primary **negative feedback** is **at the pituitary,** negative feedback regulation also occurs at the hypothalamus by blocking TRH release.
2. T_3 and free T_4 negatively feedback and inhibit the synthesis of TRH by hypothalamic neurons and TSH release from the anterior pituitary.
3. **Somatostatin** and **dopamine** act to inhibit secretion by blocking TSH secretion.

E. Iodide and Thyroid Hormone Homeostasis

1. Iodide mediates thyroid hormone biosynthesis and release.
2. An excess or deficiency of iodide intake can alter thyroid function.
3. Decreased iodide in the diet decreases T_4 and T_3 secretions. This increases TSH due to lack of feedback. Increased TSH levels cause hypertrophy and hyperplasia of the thyroid (goiter).
4. With decreased iodine availability, the thyroid compensates, secreting more of the active hormone, T_3.
5. High iodide intake will decrease both thyroid hormone biosynthesis (**Wolff-Chaikoff** effect) and release.
6. Radioactive iodide is used for diagnostic studies, for example, to evaluate increased uptake in hyperthyroidism (excess thyroid hormone secretion).
7. Ablation of the thyroid in hyperthyroid individuals can be done with ^{131}I instead of surgery.

F. Physiologic Actions of Thyroid Hormones

1. T_4 and T_3 enter cells by passive diffusion, and act via nuclear receptors in target tissues.
2. The **primary effect of the thyroid hormones** is to increase the basal metabolic rate (calorigenic action) in all tissues of the body except brain, testes, and spleen.
   a. Thyroid hormones increase **basal metabolic rate** (**BMR**) by either **heat production** or **oxygen consumption.**
      (1) Both respiration and cardiac output are increased in order to supply tissues with increased O_2.
      (2) If BMR increases, and adequate fuel (increased food intake) is not provided, catabolism results and weight is lost. (Hyperthyroid individuals often lose weight, and hypothyroid individuals often gain weight.)
      (3) The **generalized muscle wasting** in the presence of high concentrations of the thyroid hormones is associated with muscle weakness and fatigability.
      (4) High levels of thyroid hormones also cause **increased excretion of Ca^{2+}** and **PO_4^{3-}** as well as decreased bone mass, and occasionally pathologic fractures, in elderly women.
      (5) Thyroid hormones **stimulate heat production** (**thermogenesis**).

(*a*) Hyperthyroid individuals exhibit peripheral vasodilation and sweating.
(*b*) Hypothyroid individuals exhibit peripheral vasoconstriction and intolerance to cold.

3. Thyroid hormones are necessary for **normal growth and development.**
   a. Individuals who are hypothyroid from birth are dwarfed and mentally retarded. This condition is known as **cretinism** and is seen in regions of endemic iodine deficiency.
   b. If thyroid hormone replacement is not begun by the end of the first month after birth, the neurologic defects causing mental retardation cannot be reversed.
4. Many thyroid hormone actions are due to a synergistic interaction with the sympathetic nervous system (eg, thermogenesis, lipolysis, glycogenolysis, gluconeogenesis). Adrenergic blockade attenuates many of the cardiovascular and nervous system manifestations (eg, tremor) of hyperthyroidism.

HYPERTHYROIDISM AND HYPOTHYROIDISM

- ***Hyperthyroidism***
  –***Graves disease*** *is a female-dominant autoimmune disease and is the most common cause of hyperthyroidism.*
  –*Graves disease is caused by an abnormal immunoglobulin (TSI) that binds to and activates the TSH receptor.*
  –*Normal feedback mechanisms are altered in Graves disease; both T_3 suppression and TRH feedback response are impaired.*
  –***Clinical features*** *of hyperthyroidism include sinus tachycardia, nervousness, weight loss, heat intolerance, diarrhea, and* ***goiter*** *formation.*
  –*A minority of patients with Graves' disease develop* ***exophthalmos*** *(protruding eyes).*
- ***Hypothyroidism***
  –*Hypothyroidism in the United States is most commonly caused by* ***Hashimoto thyroiditis,*** *an autoimmune disease that produces antithyroid antibodies seen primarily in women.*
  –***Clinical features*** *include weakness, coarse skin, cold intolerance, weight gain, peripheral edema, constipation and a painless goiter.*

IX. Male Reproductive Hormones

A. Fetal Life

1. In normal males, the testis determining gene (TDF) is located on the Y chromosome. The TDF is a single gene called SRY that stimulates the medullary portion of the indifferent gonad to develop into a testis.
2. **Wolffian and Müllerian ducts** are initially present in both male and female fetuses. In the absence of hormonal input (ie, in the normal female fetus), female internal and external structures develop (ie, Müllerian ducts develop).
3. Normal male development requires the presence of three hormones: **testosterone, dihydrotestosterone** (DHT), and the **Müllerian-inhibiting factor** (**MIH**).
   a. The testis makes androgens and MIH necessary for the male pattern of sexual differentiation.
   b. **Sertoli cells** secrete MIH, which causes regression of the Müllerian ducts.
   c. Human chorionic gonadotropin (hCG) and LH stimulate Leydig cells to secrete testosterone, which stimulates Wolffian duct development.

d. In certain tissues (eg, prostate), **5-α-reductase** converts testosterone to **DHT,** which stimulates the development of organs from the urogenital sinus and genital tubercle.

4. In the absence of MIH, the Müllerian ducts develop, which differentiate into female internal structures.
5. Wolffian ducts differentiate into the majority of male internal structures, namely, epididymis, vasa deferentia, and seminal vesicles.
6. The urogenital sinus and genital tubercle differentiate into the scrotum, penis, and prostate gland.

5-α-REDUCTASE DEFICIENCY AND ANDROGEN INSENSITIVITY

- ***5-α-Reductase Deficiency***
  –Although individuals with 5-α-reductase deficiency have testes and Wolffian duct development, external genitalia are female due to a lack of DHT.
  –Large quantities of testosterone at puberty are often able to stimulate sufficient DHT to cause some virilization.
- ***Androgen Insensitivity*** *(Testicular Feminization)*
  –An X linked disorder where individuals have testes but lack androgen receptors and Wolffian duct structures.
  –Because of the absence of androgen effects, the affected person has female external genitalia with a blind vagina and rudimentary uterus, and develops as a female.

B. Puberty

1. The Hypothalamus secretes GnRH which acts on gonadotrophs in the anterior pituitary. GnRH is secreted in episodic bursts, which causes pulsatile release of FSH and LH from the anterior pituitary.
2. **GnRH levels are low during childhood but at puberty,** the **amplitude of the LH pulses becomes greater,** particularly during sleep, driving the mean level of LH higher.
3. The increased LH **stimulates the Leydig (Interstitial) cells** to secrete testosterone.
4. During pubertal development, the sensitivity of the gonadotrophs to feedback inhibition by testosterone decreases.

C. Aging Adult

1. **Testosterone secretion gradually decreases** with age. However, there is no abrupt decrease in testosterone secretion in men that parallels the abrupt decrease in estrogen secretion at menopause.
2. The decline of testosterone with aging is frequently not accompanied by an increase in LH secretion indicating some hypothalamic-pituitary dysfunction.
3. Adipose tissue, skin, and the adrenal cortex also produce testosterone, other androgens, and estrogen.

D. Major Cell Types of the Testis

1. LH receptors are located on the cell membranes of the **interstitial cells of Leydig.** LH stimulation results in increased synthesis and secretion of testosterone.
2. Blood testosterone provides a negative feedback signal to both the hypothalamus and the anterior pituitary to regulate LH secretion.
3. Much of the testosterone synthesized by the Leydig cells diffuses into adjacent Sertoli cells.

4. An **androgen-binding protein** (**ABP**) synthesized by the Sertoli cells and secreted into the lumen of the seminiferous tubules helps maintain a high local concentration of testosterone. **FSH** increases the synthesis of this protein.
5. **Testosterone acts** on target organs by binding to a nuclear receptor.
6. FSH stimulates Sertoli cells to synthesize products needed by both Leydig cells and developing spermatogonia.
7. FSH causes the Sertoli cell to secrete **inhibins that regulate FSH secretion.**
8. FSH occupies receptors located on the plasma membrane of Sertoli cells to increase the production of proteins such as ABP. Both FSH and testosterone are required for normal spermatogenesis.
9. Sertoli cells are the source of **MIH.**
10. Sertoli cells also secrete **aromatase,** which converts androgens to estrogens, and Sertoli cell tumors are associated with feminization in animals.

E. Anabolic Actions of Androgens

1. Androgens increase GH secretion, which drives IGF-1 to increase long bone growth, stimulating a growth spurt.
   a. Thus, androgens stimulate the growth of long bones and are responsible for the greater average height of men compared to women.
2. Near the end of puberty, androgens promote the closure of the epiphyseal plates of long bones stopping growth.
3. Protein synthesis is stimulated in muscle, causing the larger muscle mass in men as compared with women.
4. Erythropoietin secretion is stimulated by the kidneys and increases red blood cell production.
5. Stimulation of sebaceous gland secretion.

F. Androgenic Effects of Androgens

1. Androgens induce development of male accessory reproductive organs and maintenance of their secretions.
2. They increase development of male secondary sex characteristics including the growth, development and function of male internal and external genitalia.
3. They are required for behavioral effects including libido and potency.
4. They regulate differentiation of male internal and external genitalia in the fetus.
5. They are required for the initiation and maintenance of spermatogenesis.

G. Regulation of Testicular Function

1. The pulsatile release of **GnRH** from the hypothalamus stimulates FSH and LH release from the anterior pituitary.
2. **FSH** acts on Sertoli cells to regulate **spermatogenesis,** and Sertoli cells secrete inhibin, a protein that regulates FSH secretion from the anterior pituitary by negative feedback.
3. **LH stimulates** Leydig cell **testosterone secretion.** Testosterone inhibits LH by inhibiting release of GnRH from the hypothalamus and LH release from the anterior pituitary.

H. Regulation of Spermatogenesis

1. Spermatogenesis ceases at temperatures found in the abdominal cavity. If testes fail to descend (**cryptorchidism**)by birth, infertility results if this condition is not corrected early in life.
2. The scrotum provides an environment 4°C cooler than the abdominal cavity via a countercurrent heat exchanger located in the spermatic cord and the

movement of the testes toward and away from the body by cremasteric muscles that surround the testes.

3. The Sertoli cells support spermatogenesis by regulating the metabolism of seminiferous tubules.
4. Once produced in the testis, sperm maturation occurs in the epididymis.

I. Male Sexual Response

1. **Erection** is primarily under parasympathetic control. Efferent parasympathetic fibers and nonadrenergic noncholinergic fibers that release **nitric oxide** (**NO**) mediate erection.
2. An important neurocrine mediators of erection is **NO** which increases cGMP levels in vascular smooth muscle cells.
   a. Erectile dysfunction is now treated with drugs like sildenafil that prevent cGMP breakdown to maintain erectile function.
3. **Emission** is the movement of semen from the epididymis, vas deferens, seminal vesicles, and prostate to the ejaculatory ducts and is primarily under sympathetic control.
4. A sympathetic adrenergic-mediated contraction of the internal sphincter of the bladder prevents retrograde ejaculation of semen into the bladder. Destruction of this sphincter by prostatectomy often results in retrograde ejaculation.
5. **Ejaculation** is caused by the rhythmic contraction of the bulbospongiosus and ischiocavernous muscles of the urogenital diaphragm.

MALE PATHOPHYSIOLOGY

- ***Benign Prostatic Hyperplasia*** *(BPH)*
  –In elderly men, growth of the medial lobe of the prostate occludes the urethra, leading to urinary retention.
  –BPH is androgen dependent, primarily on DHT.
  –Treatment involves surgery or 5-α-reductase inhibitors to decrease prostate enlargement.
- ***Klinefelter Syndrome***
  –This syndrome occurs with variable expressivity in XXY individuals.
  –Individuals affected by this syndrome are tall, with small testes and gynecomastia (male breast enlargement).
  –Low testosterone levels lead to increased gonadotropin levels.
  –Seminiferous tubule development is abnormal; ***azoospermia*** *(lack of viable sperm) occurs.*

X. Female Reproductive Hormones

A. Fetal Life

1. No SRY antigen is produced; therefore, the indifferent gonad develops into an ovary.
2. Lack of MIH allows the Müllerian ducts to develop into the uterus and fallopian tubes.
3. Lack of testosterone causes Wolffian ducts to degenerate and prevents development of male structures.

B. Synthesis of Estrogens

1. **Theca cells** are the major sources of **17-α-hydroxyprogesterone** and of **androstenedione** (the principal androgen produced by the ovary). **Granulosa cells** are the major source of estradiol (E_2). LH stimulates progesterone synthesis from pregnenolone.

2. Significant amounts of estrogens are produced by the peripheral aromatization of androgens. In human males, the peripheral aromatization of testosterone to $\mathbf{E_2}$ accounts for 80% of the production rate of estrogens.
3. In females, as much as 50% of the E_2 produced during pregnancy comes from the aromatization of the adrenal androgen **DHEA sulfate.**
4. The conversion of androstenedione to **estrone** ($\mathbf{E_1}$), is the major source of estrogens in postmenopausal women. Aromatase activity is present in adipose cells and in liver, skin, and other tissues.

C. **Physiologic Effects of Estrogen**

1. Estrogen is responsible for development of female secondary sex characteristics, including
   a. Narrow shoulders
   b. Broad hips and wider carrying angle
   c. Divergent arms
   d. Convergent thighs and wider pelvic inlet
2. Estrogen has the following **endocrine organ effects:**
   a. It increases keratinization of the vaginal lining for protection.
   b. It is responsible for a profuse watery cervical secretion that can be stretched into long threads (**spinnbarkheit**) used clinically to indicate that ovulation is imminent.
   c. It enhances the growth of the endometrial (secretory) layer, making it 3–4 times thicker.
      (1) Increases actin and myosin of myometrium and fallopian tubes to promote spontaneous contractions to facilitate sperm transport.
   d. **Increases** the number of LH receptors on ovarian follicles, and follicular growth increases.
   e. Increases ductal development (ie, number and size) in the breast.
3. Estrogen has the following **metabolic effects:**
   a. It lowers blood levels of cholesterol and inhibits atherogenesis in animals. High doses, however, promote thrombosis.
   b. It increases Ca^{2+} retention. Thus, at menopause, Ca^{2+} loss occurs. Estrogen also promotes pubertal growth and then closure of epiphyseal plates.
   c. Estrogen increases the number of steroid-binding proteins (eg, thyroglobulin) synthesized by the liver.

D. **Progesterone**

1. **Progesterone** has the following **endocrine effects:**
   a. It increases leukocyte infiltration in vaginal epithelium.
   b. It produces a thick cervical mucous resistant to the penetration of spermatozoa (natural fertilization barrier).
   c. It decreases uterine spontaneous activity and sensitivity to oxytocin, while increasing endometrial gland secretion.
   d. It stimulates lobular alveolar gland growth in the breast.
2. **Progesterone** has the following **metabolic effects:**
   a. It induces natriuresis (Na^+ loss from the kidney) by competing for aldosterone receptors.
   b. It makes the respiratory system more sensitive to CO_2, thereby increasing respiratory rate.
   c. It increases the basal body temperature by about 1°C.

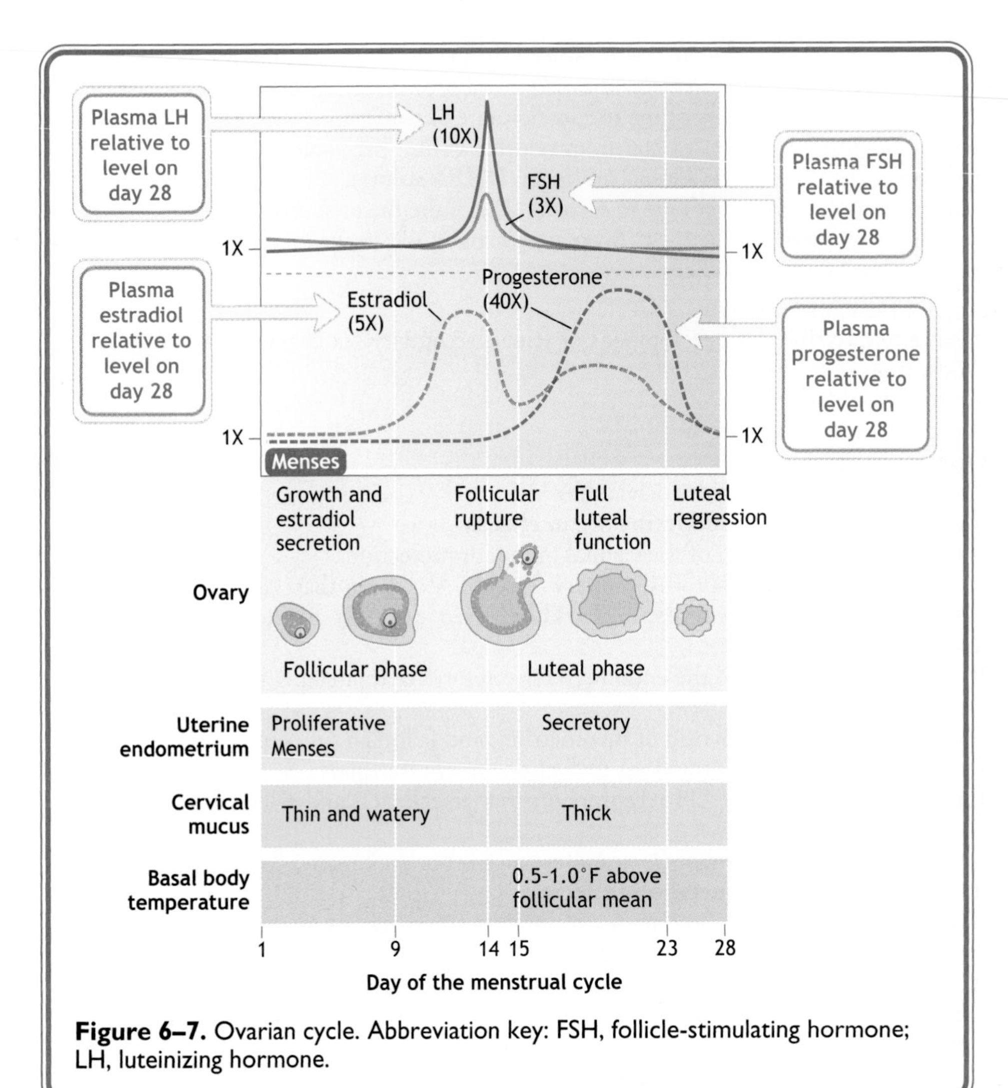

Figure 6–7. Ovarian cycle. Abbreviation key: FSH, follicle-stimulating hormone; LH, luteinizing hormone.

E. Ovarian Cycle: Hormonal Regulation of Oogenesis (Figure 6–7)

1. The female reproductive system has cyclic variation, whereas the male system has tonic and constant production of hormones.
   a. Puberty marks the transition to cyclic adult reproductive function.
2. Cyclic activity in ovarian steroid secretion produces changes that are reflected by sloughing off of the endometrial lining approximately every 28 days. Menses (*month* in Latin) is due to ovarian activity and, therefore, is named the **ovarian cycle.**
3. The cycle begins with 15–30 follicles developing due to FSH stimulation.
4. On day 6, one follicle begins to produce antifollicular compounds, which lead to atresia of other follicles.
5. The dominant follicle produces estrogen, which decreases FSH and LH by negative feedback.

6. Near midcycle, about 48 hours prior to ovulation, a huge increase in estrogen production results in a **positive feedback** that causes a surge of gonadotropins (primarily LH), leading to rupture of the follicle and ovulation 16–24 hours later.
7. After ovulation, the follicular cavity fills with yellowish luteal cells due to luteinization of the theca and granulosa cells (corpus luteum) resulting from LH stimulation. The corpus luteum produces progesterone and E_2 for 8–9 days.
8. The increased plasma levels of E_2 and progesterone negatively feed back to keep FSH and LH levels low.
9. If fertilization does not occur, the corpus luteum then regresses (**luteolysis**).
10. Days 1–14 of the ovarian cycle are called the **follicular phase** (preovulatory). After ovulation, the **luteal phase** (postovulatory) begins with corpus luteum functioning. The follicular phase is the most variable, whereas the **luteal phase** is always 14 days. Thus, the day of ovulation can be estimated by subtracting 14 days from the total length of the menstrual cycle.
11. LH levels continue to decline during the luteal phase, whereas FSH levels rise during the late luteal phase due to a decline in E_2 and progesterone secretion and removal of negative feedback effects on the hypothalamo-hypophyseal complex.

F. Uterine Cycle (Figure 6–8)

1. The hormonal pattern in the uterus reflects changes in ovarian function because it is the target organ of E_2 and progesterone.
2. **Menstruation,** or sloughing of the functional layer of endometrium, occurs during **days 1–5.**
3. **Days 5–14** represent the **proliferative phase,** during which the endometrium shed at menstruation is restored.

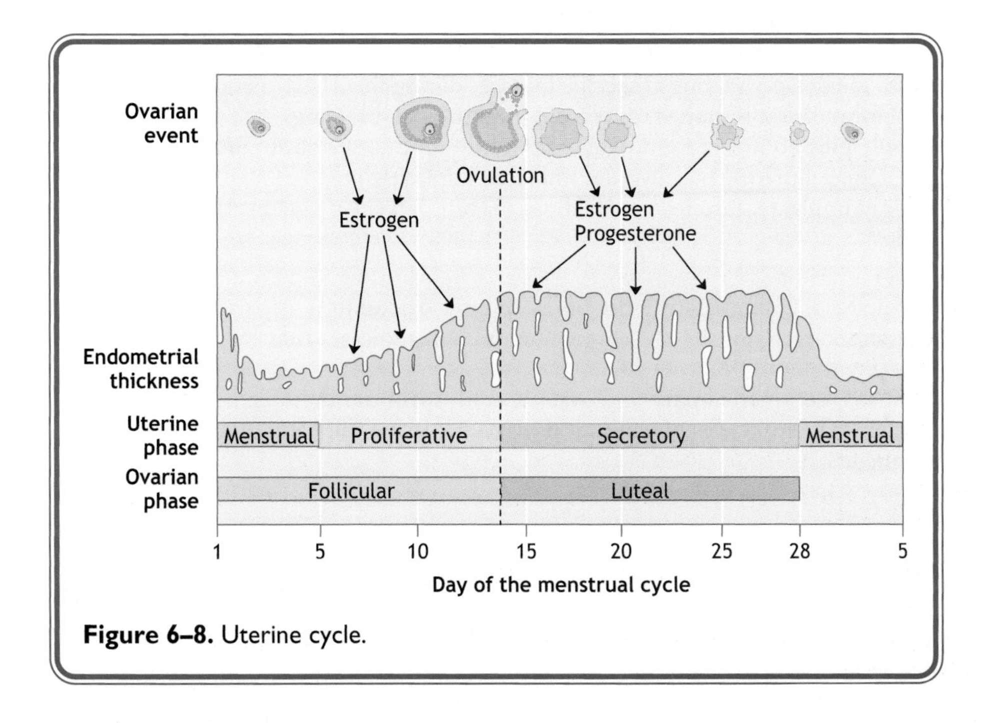

Figure 6–8. Uterine cycle.

a. The endometrium of the uterus grows and develops.
b. The endometrial lining increases in thickness three to four times; cells become edematous and more highly vascularized at ovulation.
c. The length of this phase is highly variable.

4. **Days 14–28** represent the **secretory phase,** during which the uterus prepares for implantation of the fertilized ovum.
   a. After ovulation, glands become coiled and endometrial secretions increase.
   b. This phase always lasts 14 days.
5. If fertilization does not occur, hormonal support of the endometrium ceases, the endometrium regresses, and a new cycle begins with first day of menstruation.

G. Hormonal Control of the Ovary

1. Neurons in the hypothalamus release GnRH in a pulsatile fashion which stimulates production of FSH and LH by the anterior pituitary.
2. FSH and LH act on the ovary to stimulate follicular development and estrogen production.
3. In the early follicular phase, low levels of estrogen negatively feeds back to the hypothalamus and anterior pituitary to inhibit LH production.
4. Inhibin, a peptide hormone, produced by the granulosa cells of the follicle exerts negative feedback to inhibit FSH secretion.
5. At midcycle, both negative feedback and positive feedback are present. High levels of estrogen dominate for at least 36 hours to positively feed back to the pituitary, causing the LH and FSH surge that results in ovulation.
6. Following ovulation, when circulating levels of progesterone and estrogen are high, the positive feedback of estrogen is inhibited.
7. Most oral contraceptives contain a synthetic progestin and estrogen, which combine to prevent gonadotropin release.
8. Oral contraceptives are administered for 21 days, then withdrawn for 5–7 days to permit menstrual flow to occur.
9. Multiphase oral contraceptives contain different amounts of estrogen and progesterone with the dosages varying at specific intervals during the 21-day period. This type of formulation is thought to prevent breakthrough bleeding.
10. Progestin-only pills do not prevent ovulation but do produce cervical mucus thickening, reduced uterine motility and impaired implantation.
11. Progestin-only implants, inserted under the skin, can prevent pregnancy for up to 5 years.

H. Indicators of Ovulation

1. The approximate 1°C temperature rise observed after ovulation is associated with progesterone secretion by the corpus luteum and indicates that ovulation has occurred.
2. Cervical mucus becomes copius and watery, and **spinnbarkheit** can be formed, associated with the estrogen peak prior to ovulation indicates ovulation is imminent.

OVARIAN DISEASE

- ***Polycystic ovarian disease*** *is a syndrome characterized by multiple ovarian cysts and elevated E_2, testosterone, and LH levels, with decreased FSH levels.*
- *One proposed cause of this syndrome is very frequent GNRH pulse generation.*
- ***Treatment*** *often involves wedge resection of the ovary.*

I. Pregnancy-Associated Endocrine Changes

1. With fertilization, the corpus luteum does not regress and instead enlarges in response to **hCG** released from the syncytiotrophoblast (Figure 6–9).
2. **hCG** bridges the gap between ovarian and placental maintenance of pregnancy.
3. **hCG** is similar to LH in both structure and action. Thus, it stimulates the production of progesterone, 17-hydroxyprogesterone, and estradiol by the corpus luteum.
4. The syncytiotrophoblast also secretes **human chorionic somatomammotropin (hCS)**, also known as **human placental lactogen (hPL),** which is the maternal growth hormone of pregnancy. hCS brings about decreased maternal glucose utilization.
5. Another major estrogen of human pregnancy is **estriol,** which is used as a marker for fetal-placental health, because its production requires the shuttling of steroid molecules from the placenta to the fetus and back again.

J. Parturition

1. Pregnancy lasts a predetermined number of days for each species (270 days from fertilization and 284 days from the last menstrual period preceding conception in the human), but the factors responsible for its termination are unknown.
2. There are 100 times more oxytocin receptors in the uterus at term than there are at the onset of pregnancy, which correlates with the increased amount of estrogen at term.
3. Although the mechanism by which labor is initiated is not completely understood, a local change in the estrogen-to-progesterone ratio is thought to increase **prostaglandin release,** causing the onset of uterine contractions.
4. Once labor begins, oxytocin increases uterine contractions by acting directly on muscle cells and increasing prostaglandin production.

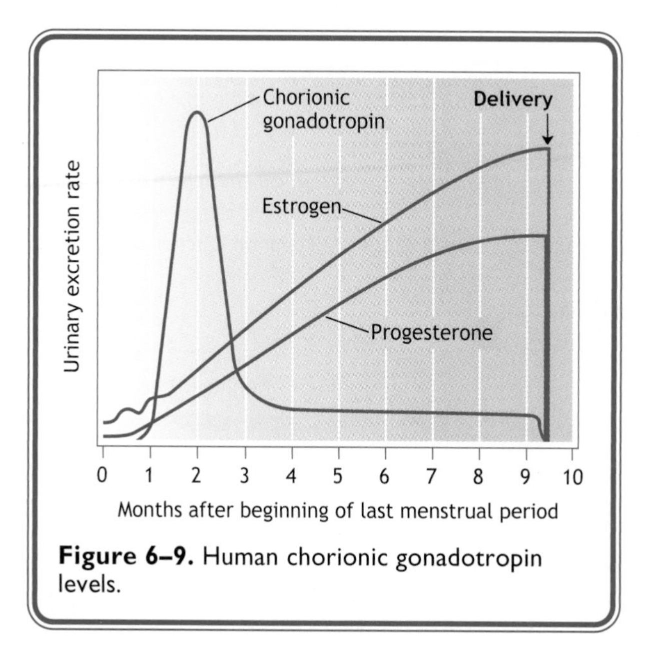

Figure 6–9. Human chorionic gonadotropin levels.

K. Hormone Alterations in Lactation

1. Estrogen and progesterone stimulate development of mammary glands during pregnancy.
2. Milk synthesis by prolactin and hPL begins in the last trimester of pregnancy.
3. Lactogenesis (milk synthesis and secretion) does not occur during pregnancy because progesterone and estradiol inhibit prolactin stimulation of milk synthesis.
4. After delivery, the absence of suckling allows prolactin levels to remain low and prevent milk formation. Nursing stimulates prolactin levels via a neuroendocrine reflex.
5. Nursing also stimulates oxytocin release, causing contraction of the alveolar myoepithelium, resulting in milk letdown.
6. Milk letdown can also be stimulated by emotional reactions to the baby and by sexual activity.
7. Maintenance of nursing suppresses ovulation because of the antigonadotrophic actions of prolactin.

L. Menopause

1. **Menopause** is the time at which the last menstrual cycle occurs. The average age at menopause has increased in recent years and is currently 52 years.
2. At menopause, only a few functioning follicles remain in the ovary. Failure of the ovary to respond to gonadotrophins results in the following endocrine changes:
   a. The predominant premenopausal estrogen is estradiol, but after menopause higher levels of estrone are found. Estrone is produced in peripheral adipose tissue by the aromatization of androstenedione.
   b. Levels of circulating gonadotrophins, particularly FSH, increase.
   c. Gradual atrophy of reproductive organs occurs, due to loss of hormonal support.
   d. Osteoporosis becomes common due to estrogen loss.
   e. Vasomotor instability and hot flashes occur.

PREMATURE OVARIAN FAILURE: TURNER SYNDROME

- *Turner Syndrome is the best known example of gonadal dysgenesis.*
- ***Clinical features** of Turner syndrome are short stature, webbed neck, shield chest, increased carrying angle, and coarctation of the aorta.*
- *The defect involves an absence of an X chromosome, leading to loss of ovarian function and the clinical features described above.*
- *It is the **most common cause of primary amenorrhea.***
- *Clinically, the ovaries are replaced by fibrous streaks.*

CLINICAL PROBLEMS

A 46-year-old woman complains of increasing irritability, hot flashes, and an increasingly irregular menstrual cycle over the past 12 months. She has had three uncomplicated pregnancies and has no other problems other than varicose veins.

1. Which of the following findings would be expected?
   A. Increased FSH
   B. Decreased LH
   C. Increased bone density
   D. Decreased total cholesterol
   E. Increased estradiol levels

A 19-year-old unmarried woman complains that she is "too hairy." She has a history of severe irregular menstrual cycles; worsening acne; and hair growth on the face, breasts, and lower abdomen. She is sexually active, uses condoms for contraception, and is otherwise well. Laboratory values revealed normal levels of testosterone, cortisol, 17-hydroxyprogesterone, and DHEA but elevated LH levels. She is, however, depressed about the new hair growth.

2. Which of the following is the most likely cause for these symptoms?
   A. Congenital adrenal hyperplasia
   B. Sertoli-Leydig cell tumor of the ovary
   C. Polycystic ovarian disease
   D. Cushing syndrome
   E. Constitutional hirsutism

A 34-year-old female patient has an 8-year history of menorrhagia (abnormal bleeding) and anemia. Symptoms of tiredness and inability to concentrate caused her to lose her job. She also complains of cold intolerance, constipation, and weight gain. Upon examination her skin is pale, cold, dry, and scaly. Laboratory results indicate low plasma T_4 levels and elevated plasma TSH levels.

3. Which of the following is the most likely diagnosis?
   A. Graves disease
   B. Multinodular toxic goiter
   C. Hypothyroidism of hypothalamic origin
   D. Primary hypothyroidism
   E. Diffuse toxic goiter

A 32-year-old previously healthy woman has a 1-year history of palpitations, sweating, heat intolerance, and intermittent diarrhea. She has lost 12 lbs despite a good appetite. Her niece had neonatal hypothyroidism. Physical examination reveals an anxious woman with a pulse rate of 120/min and blood pressure of 120/80 mm Hg. She exhibits a fine tremor of the hands and has moist, warm palms. Laboratory investigation shows elevated free T_3 and T_4 as well as increased plasma TSH levels.

4. Which of the following is the most likely diagnosis?
   A. Graves disease
   B. Multinodular toxic goiter
   C. Hypothyroidism of hypothalamic origin
   D. Primary hypothyroidism
   E. Myxedema

A 30-year-old man has a 1-week history of increased thirst (polydipsia) and increased urine volumes (polyuria). The results of a water-deprivation test reveal increased plasma osmolality (>300 mOsm/kg) and elevated urine osmolality (>120 mOsm/kg). He is currently receiving treatment for non-insulin-dependent diabetes mellitus (type II) as well as for bipolar disorder.

5. Which of the following is the most likely diagnosis?
   A. SIADH due to oral hypoglycemic agents
   B. Hyperglycemia with osmotic diuresis
   C. Nephrogenic diabetes insipidus due to lithium treatment
   D. polydipsia
   E. Central diabetes insipidus caused by panhypopituitarism

A 48-year-old farmer has a 10-month history of muscle weakness, easy bruising, backache, headache, and depression. He is a lifelong nonsmoker who has previously been healthy. His only medication is a nonsteroidal anti-inflammatory agent taken for rib pain. Upon examination, truncal obesity with a "buffalo hump," thin skin with easy bruising, and a blood pressure of 180/100 mm Hg are noted. Laboratory studies reveal elevated free cortisol with an absence of a circadian rhythm. A high-dose dexamethasone test suppressed AM cortisol levels to less than 50% of basal values.

6. Which of the following is the most likely diagnosis?
   A. Addison disease due to autoimmune destruction of the adrenal
   B. Ectopic Cushing disease due to small cell carcinoma of the lung
   C. 17-α-Hydroxylase deficiency due to a congenital defect
   D. Pituitary Cushing disease due to a pituitary adenoma
   E. Primary hyperaldosteronism from adrenal adenoma

ANSWERS

1. A is correct. The subject is experiencing early menopause, which is characterized by increased FSH trying to stimulate follicular development and maturation. Decreased LH (choice B), increased bone density (choice C), decreased total cholesterol (choice D), and increased estradiol levels (choice E) are not characteristic of menopause. Rather, the opposite findings are associated with menopause.

2. C is correct. The history of hair growth and irregular menstrual cycles points to polycystic ovarian disease, the most common cause of androgen excess and hirsutism. Patients exhibit obesity, oligomenorrhea, hirsutism, acne, and infertility. The distribution of abnormal hair growth reflects the severity of androgen excess. Congenital adrenal hyperplasia (choice A) can be excluded because it is characterized by increased 17-hydroxyprogesterone levels. Sertoli-Leydig cell tumors (choice B) are ovarian neoplasms that secrete testosterone. They occur in women between the ages of 21 and 40 years. Patients with these tumors abruptly cease having menses and exhibit extensive body hair as well as temporal hair recession, clitoral enlargement, deepening of the voice, and an

ovarian mass. Cushing syndrome (choice D) is unlikely because free cortisol levels were normal. Constitutional hirsutism (choice E) is hirsutism with no known cause. Women with this condition have regular menstrual cycles.

3. D is correct. Laboratory test results (low thyroid hormone and elevated serum TSH levels) along with physical findings of pale, cold, dry skin are consistent with primary hypothyroidism. Graves disease (choice A) is an autoimmune disorder characterized by production of antibodies to the TSH receptor. Hyperthyroidism results when thyroid-stimulating immunoglobulins act as agonists on the TSH receptor, stimulating thyroid hormone synthesis and secretion. Multinodular toxic goiter (choice B) is most commonly seen in the elderly and results in hyperthyroidism, not hypothyroidism. Hypothyroidism of hypothalamic origin (choice C) is an uncommon cause of hypothyroidism and is due to central (pituitary/hypothalamic) defects that result in decreased TSH levels, not increased TSH levels. Diffuse toxic goiter (choice E) is another name for Graves disease and is characterized by hyperthyroidism due to an overactive thyroid gland, not hypothyroidism.

4. A is correct. In most cases of Graves hyperthyroidism, an IgG antibody to TSH receptors known to stimulate thyroid cells is present in the patient's serum. Multinodular toxic goiter (choice B) occurs primarily in elderly patients. Hypothyroidism of hypothalamic origin (choice C); primary hypothyroidism (choice D); and myxedema, which is prolonged hypothyroidism (choice E), are incorrect because the case describes hyperthyroidism, not hypothyroidism.

5. C is correct. Lithium treatment decreases the effects of ADH on the kidney, resulting in polyuria and polydipsia from volume depletion. SIADH (choice A) is associated with inappropriate ADH secretion, leading to hyponatremia and defective excretion of a water load, which is contrary to the symptoms described in the case. Hyperglycemia with osmotic diuresis (choice B) causes a loss of water that results in increased ADH release as the hypothalamus senses hypovolemia and secretes ADH to retain water to maintain intravascular volume. Because the patient has a long history of diabetes, symptoms would not develop in 1-week's time. In a patient with psychogenic polydipsia (choice D), the urine osmolality would increase more than the plasma osmolality after the water restriction test, indicating that antidiuresis is occurring. Central diabetes insipidus with panhypopituitarism (choice E) is not correct because a person with panhypopituitarism would exhibit multiple pituitary deficiencies, including growth failure and hypogonadism, which were not mentioned in this case.

6. D is correct. Pituitary Cushing disease is due to a benign pituitary adenoma that secretes excessive ACTH, producing clinical symptoms of excessive cortisol levels. Addison disease (choice A) is adrenal insufficiency due to autoimmune destruction of the adrenal, leading to symptoms of cortisol deficiency, not cortisol excess. Ectopic Cushing disease is due to a malignant tumor of the lung (choice B) that produces excess ACTH. Patients exhibit signs of obvious metastatic tumor such as weight loss, hypertension, hypokalemia, and hyperpigmentation. Weight loss and hyperpigmentation were not mentioned in this case. Congenital 17-α-hydroxylase deficiency (choice C) causes a failure of androgen and estrogen formation and, therefore, presents as a female phenotype with absent secondary sex characteristics. Primary hyperaldosteronism (choice E) due to an adrenal adenoma is the most common cause of primary aldosteronism. Common clinical manifestations include hypertension, hypovolemia, hypomagnesemia, and metabolic alkalosis. This case describes symptoms of elevated cortisol levels not elevated aldosterone levels.

CHAPTER 7
NEUROPHYSIOLOGY

I. Autonomic Nervous System

A. Organization

1. The autonomic nervous system (ANS) has three divisions: **sympathetic, parasympathetic, and enteric.**
2. Synapses between neurons are made in autonomic ganglia.
   a. **Parasympathetic ganglia** are located in or near the effector organs.
   b. **Sympathetic ganglia** are located in the paravertebral chain (along the vertebral column).
3. Each division has two neurons in the peripheral distribution of the motor innervation (Figure 7–1).
   a. A **preganglionic neuron** has its cell body in the **central nervous system** (**CNS**) in columns of cells in the brain stem and spinal cord.
      (1) Preganglionic sympathetic neurons originate in the thoracolumbar spinal cord segments T1–L3.
      (2) Preganglionic parasympathetic neurons originate in the nuclei of **cranial nerves** (**CNs**) **III, VII, IX, and X** and in **spinal cord segments S1–S4.**
   b. A **postganglionic neuron** has its cell body in a ganglion in the **peripheral nervous system** (**PNS**) between the CNS and target cells.
4. The effect of increased activity in either system is **excitatory** in some target organs and inhibitory in others.
5. The **adrenal medulla** is a specialized ganglion of the **sympathetic nervous system.**
   a. **Preganglionic fibers** synapse directly on **chromaffin cells,** which act like postganglionic cell bodies.
   b. **Chromaffin cells** contain **phenylethanolamine N-methyltransferase,** which converts **norepinephrine** to **epinephrine.**
   c. Thus, the adrenal medulla secretes **80% epinephrine** and **20% norepinephrine.**
6. The **enteric division** of the ANS is a system of afferent neurons, interneurons, and motor neurons that form **plexuses** that surround the GI tract.
   a. The enteric division is a self-contained nervous system that receives both sympathetic and parasympathetic input.

B. Neurotransmitters

1. All preganglionic neurons release acetylcholine and stimulate **nicotinic** receptors on postganglionic neurons.

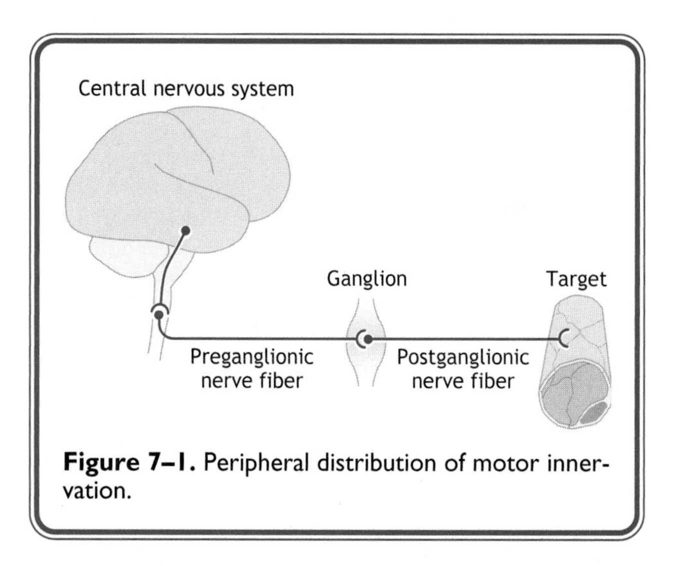

Figure 7–1. Peripheral distribution of motor innervation.

2. **All postganglionic parasympathetic neurons** release **acetylcholine and stimulate muscarinic receptors on visceral targets.**
3. Most postganglionic sympathetic neurons release **norepinephrine onto visceral targets.**
4. **Peptidergic neurons** in the parasympathetic nervous system release neurocrine peptides such as **vasoactive inhibitory peptide.**

C. Adrenergic Receptors (Table 7–1)

1. **α_1 Receptors are the dominant α receptor subtype on the** postsynaptic target cell membrane.
   a. They are located on **vascular smooth muscle, gastrointestinal** (**GI**) and **bladder sphincters,** and **radial muscle of the eye.**
   b. They are **excitatory** and produce **contraction through activation of phospholipase C,** leading to the formation of **inositol triphosphate** (**IP_3**)and an increase in **intracellular Ca^{2+}.**
2. **α_2-Receptors are the dominant α-receptor type on the** presynaptic side of adrenergic nerve terminals.
   a. They are located in **presynaptic nerve terminals, platelets, fat cells,** and **walls of the gut.**
   b. They are **inhibitory** and produce **relaxation by inhibition of adenylate cyclase and by decreasing cyclic AMP** (**cAMP**).
3. **β_1-Receptors, which are found mostly in** cardiac muscle cells, are **excitatory** and produce **increased heart rate and contractility** by activation of adenylate cyclase and **by increasing cAMP levels.**
4. **β_2-Receptors are found in** smooth muscle and in **secretory effectors.**
   a. They are **inhibitory** and produce **relaxation** (eg, **dilation of bronchioles**) by **increasing cAMP levels.**
   b. They are more sensitive to epinephrine than norepinephrine.
5. **β_3-Receptors have limited distribution.**
   a. Low levels are present in **adipose tissue** and the **GI tract.**
   b. They are **excitatory** and **stimulate lipolysis and GI motility by increasing cAMP levels.**

Table 7–1. Effects of the autonomic nervous system on organ systems.

Organ	Sympathetic Action	Sympathetic Receptor	Parasympathetic Action
Heart	↑ heart rate ↑ contractility ↑ AV node conduction	β_1 β_1 β_1	↓ heart rate ↓ contractility (atria) ↓ AV node conduction
Vascular smooth muscle	Constricts blood vessels in skin; splanchnic Dilates blood vessels in skeletal muscle	α_1 β_2	— —
Gastrointestinal tract	↓ Motility Constricts sphincters	α_2, β_2 α_1	↑ Motility Relaxes sphincters
Bronchioles	Dilates bronchiolar smooth muscle	β_2	Constricts bronchiolar smooth muscle
Male sex organs	Ejaculation	α_2	Erection
Bladder	Relaxes bladder wall Constricts sphincter	β_2 α_1	Contracts bladder wall Relaxes sphincter
Sweat glands	↑ sweating	Muscarinic (sympathetic cholinergic)	—
Kidney	↑ renin secretion	β_1	—
Fat cells	↑ lipolysis	β_1	—

D. Cholinergic Receptors (see Table 7–1)

1. **Nicotinic receptors** are postsynaptic receptors in **ganglia** located in the **heart, smooth muscle,** and **exocrine glands.**
   a. They are **activated by low concentrations of acetylcholine** or **nicotine** and are **inhibited by ganglionic blockers** (eg, **hexamethonium**) and **high concentrations of acetylcholine.**
   b. They **produce excitation by acting as nonspecific cation channels, increasing the influx of Na^+ and K^+ down their electrochemical gradients.**
2. **Muscarinic receptors** are responsible for most parasympathetic postsynaptic effects and are located in the **heart, smooth muscle,** and **glands.**
   a. They are **activated by acetylcholine and muscarine** and are **inhibited by atropine.**
   b. They are **inhibitory in the heart** (eg, decreasing heart rate) and **excitatory in smooth muscle and glands** (eg, increasing secretion).
   c. They produce **inhibition by decreasing cAMP,** which leads to opening of K^+ channels, and **excitement by increasing** IP_3-mediated **release of Ca^{2+} from intracellular stores.**

E. Central Coordination

1. The **medulla** and **pons** areas of the midbrain are the most significant sites for central autonomic regulation of individual variables such as digestion, respiration, heart rate, and blood pressure.
2. **Electrical information** reaches the medulla-pons area primarily **through the nucleus tractus solitarius,** and **chemical information** reaches it mostly **through the area postrema.**
3. The **medulla-pons** area contains the **vasomotor, respiratory, pneumotaxic,** and **swallowing** and **vomiting centers.**
4. **Efferent autonomic information leaves** central nuclei in **sympathetic and parasympathetic tracts.**
5. **Sympathetic efferents descend in** the intermediolateral column of **the spinal cord** and are **transferred** from there **to sympathetic preganglionic neurons.**
6. **Parasympathetic efferents leave** the central nuclei primarily by way of the **vagus nerves. Fibers of the sacral region** of the spinal cord **descend in the mediolateral area.**
7. The **nucleus tractus solitarius** and **area postrema have extensive communications** with nuclei that generate efferent information: the **nucleus ambiguus,** the **dorsal motor nucleus,** and the **rostral** and **caudal ventrolateral medulla nuclei.**
8. **Overlap in ANS control systems can occur and be pathologic.**
   a. **Many neurons in the brain stem and spinal cord have firing patterns that are modulated by heartbeat and respiratory activity.**
      (1) **Sinus arrhythmia can occur because of an exaggerated effect of respiratory activity on heart rate.**

II. Sensory System

A. Key Concepts

1. **Sensory transduction** is the process of transforming properties of the external and internal environments into nerve impulses that can be transmitted to the brain.
2. Environmental signals that can be detected include pressure, light, sound, temperature, and chemicals.
3. These signals are detected through four categories of sensory receptors (Table 7–2).
   a. There are four types of **mechanoreceptors:**
      (1) **Cochlear hair cells** are found in the ear.
      (2) **Golgi tendon organs** and **joint receptors** are found in muscle and joints.
      (3) **Pacinian corpuscles and Meissner's corpuscles** are found in skin and viscera.
      (4) **Arterial baroreceptors** are found in the cardiovascular system.
   b. There are two types of **chemoreceptors:**
      (1) **Taste receptors** are modified epithelial cells, and **olfactory** receptors are neurons.
      (2) **Pain receptors,** hypothalamic **osmoreceptors,** and **carotid body O_2 receptors** are found in the skin and viscera.
   c. There are two types of **photoreceptors: rods** and **cones of the retina.**
      (1) **Color vision depends on the spectral sensitivities of the three types of cones.**

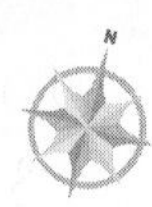

Table 7–2. Sensory receptors.

Modality	Receptor Type	Afferent Nerve Fiber Type and Conduction Velocity
Touch	Rapidly adapting mechanoreceptors (eg, hair follicle receptors) Bare nerve endings (eg, pacinian corpuscles)	Aβ 6–12 μm diameter 33–75 m/s
Touch and pressure	Slowly adapting mechanoreceptors (eg, Merkel's disks, Ruffini corpuscles)	Aβ 6–12 μm diameter 33–75 m/s
	Bare nerve endings	Aδ 1–5 μm diameter 5–30 m/s
Vibration	Meissner's corpuscles Pacinian corpuscles	Aβ 6–12 μm diameter 33–75 m/s
Temperature	Cold receptors	Aδ 1–5 μm diameter 5–30 m/s
	Warm receptors	C-fibers 0.2–1.5 μm diameter 0.5–2.0 m/s
Pain	Bare nerve endings (fast, pricking pain)	Aδ 1–5 μm diameter 5–30 m/s
	Bare nerve endings (slow, burning pain; itch)	C-fibers 0.2–1.5 μm diameter 0.5–2.0 m/s

d. There are two types of **thermoreceptors:**
   (1) **Warm and cold receptors** are found in the skin.
   (2) **Temperature-sensing hypothalamic neurons** are found in the CNS.

4. Each type of receptor is best excited by a specific type of stimulus known as its **adequate stimulus.**

5. Receptors send their information to the CNS via **afferent nerve fibers.**

6. Each afferent nerve fiber responds to a stimulus over a certain area and intensity known as its **receptive field.** If the **firing rate** of the sensory neuron **is increased,** the **receptive field is excitatory,** and vice versa.

7. **Sensory transduction leads to a change in membrane potential called a receptor potential** (Figure 7–2).

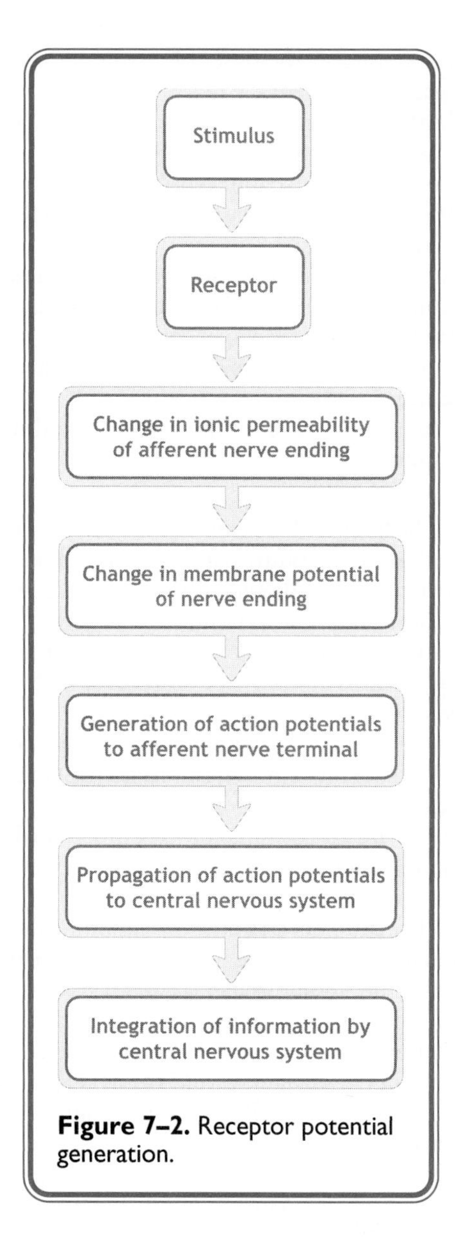

Figure 7–2. Receptor potential generation.

a. Stimulation of most sensory receptors causes cation-permeable ion channels to open, leading to depolarization of the fiber and generation of the receptor potential.
b. When threshold is reached, an action potential is generated and then transmitted by afferent nerve fibers to the CNS.
c. The **magnitude** and **duration** of a **receptor potential determines** the **number** and **frequency of action potentials generated.**

8. **Adaptation** is the fall in action potential frequency with time despite continued intensity of the stimulus.

a. **Rapidly adapting receptors** respond to the onset of a stimulus with a few action potentials and then become quiescent.

b. **Slowly adapting** or **nonadapting receptors** maintain a steady flow of action potentials as long as the stimulus is maintained.

9. Sensory pathways to the cerebral cortex involve four types of neurons:

a. **First-order neurons** are **primary afferent neurons** in the **dorsal root** or **spinal cord ganglia** that receive the transduced signal.

b. **Second-order neurons** are in the spinal cord or brainstem and **transmit information** received from primary afferent neurons **to the thalamus,** usually crossing the midline in a relay nucleus in the spinal cord.

c. **Third-order neurons** are in relay nuclei of the thalamus and **transmit information to the cerebral cortex.**

d. **Fourth-order neurons** are in sensory areas of the cortex and **allow conscious perception** of the stimulus.

B. Somatosensory System

1. The skin is the interface between the body and the environment and has receptors that sense **touch, pressure, vibration, temperature,** and **pain** (see Table 7–2).

2. There are two sensory system pathways: the **dorsal column–medial lemniscal system** and the **anterolateral (spinothalamic) system.**

a. The **dorsal column–medial lemniscal system** carries sensory information for **discriminative touch, joint position, sense, vibratory,** and **pressure sensations** from the trunk and limbs (Figure 7–3).

(1) Primary afferent neurons are located in **dorsal root ganglion cells,** and fibers (**group II**) ascend ipsilaterally and coalesce in the **fasciculus gracilis** or **fasciculus cuneatus.**

(2) These two fasciculi form the **dorsal columns of the spinal cord** and ascend the length of the spinal cord to the **nucleus gracilis** and **nucleus cuneatus** of the medulla.

(3) From the medulla, **second-order neurons** cross the midline and ascend to the **contralateral thalamus,** where they synapse with **third-order neurons,** which ascend to the **somatosensory cortex** to synapse on **fourth-order neurons.**

b. The **anterolateral system** (**spinothalamic tract**) carries **pain, temperature,** and **crude touch sensations** from the extremities and trunk (Figure 7–4).

(1) Primarily **group III** and **IV dorsal root fibers** enter the spinal cord and **synapse in the dorsal horn.**

(2) **Second-order neurons** cross the midline to the anterolateral segment of the spinal cord and ascend to the **contralateral thalamus** via the **spinothalamic tract** to synapse with third-order neurons.

(3) **Third-order neurons** ascend to the **somatosensory cortex of the postcentral gyrus** to synapse on **fourth-order neurons.**

3. Pain sense organs are the naked nerve endings found in almost every tissue in the body.

a. **Pain perception** is associated with detection of noxious stimuli (via **nociceptors**).

b. Pain is transmitted via two fiber systems:

(1) **Fast pain fibers** (**group III fibers**) cause a sharp, localized sensation that has a rapid onset and offset.

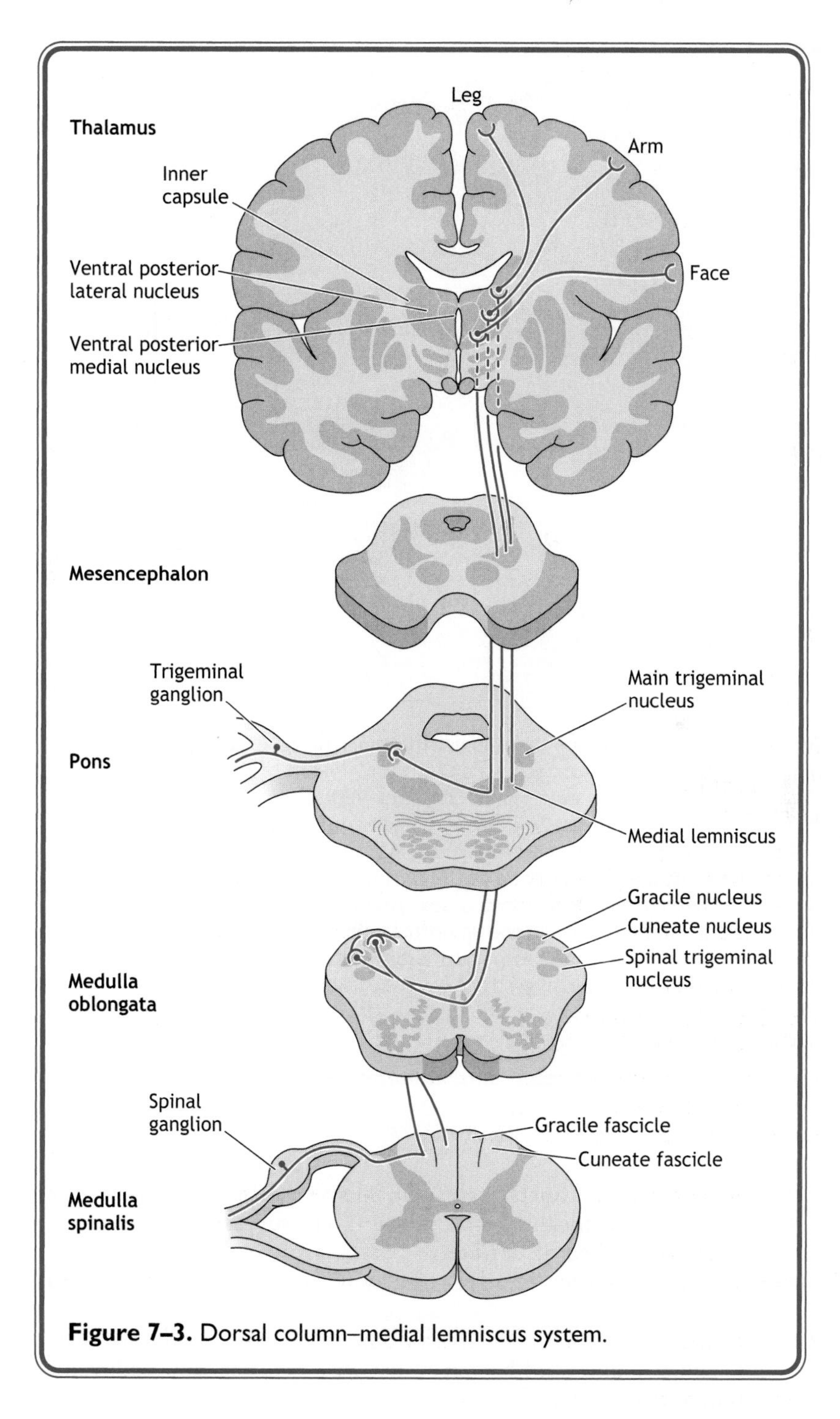

Figure 7–3. Dorsal column–medial lemniscus system.

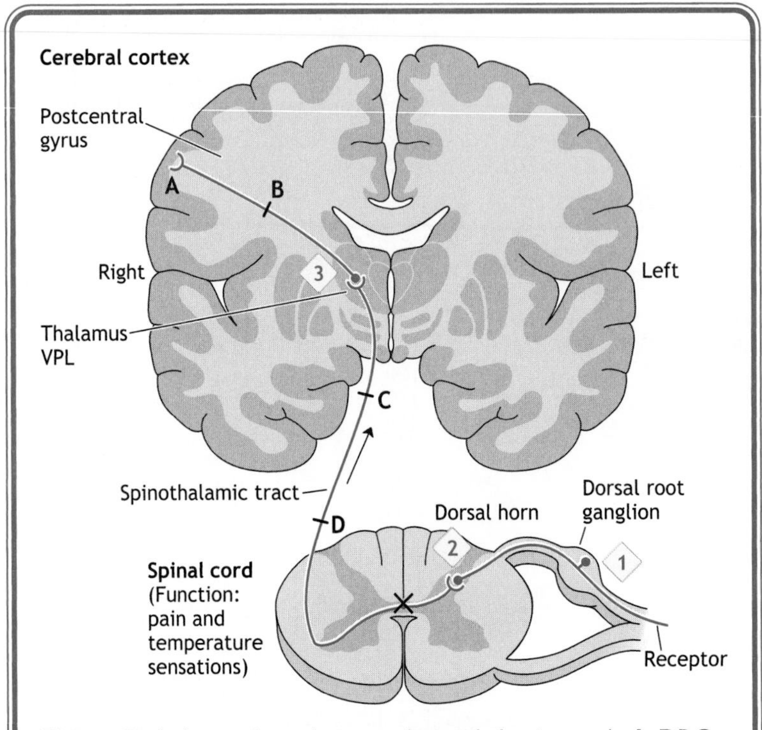

Figure 7–4. Anterolateral system (spinothalamic tract). ***1.*** DRG-dorsal root ganglion-first-order neuron. **2.** Dorsal horn of the spinal cord-second-order neuron. **3.** Thalamus-ventral posterolateral (VPL) nucleus. From the VPL nucleus, thalamocortical fibers project to the primary somatosensory area of the postcentral gyrus in the most anterior portion of the parietal lobe. Because pain and temperature information crosses soon after entering the spinal cord, unilateral lesions in the spinothalamic tract result in contralateral loss of pain and temperature. Sites of lesions producing anesthesia: postcentral gyrus (***A***), spinothalamic tract between the postcentral gyrus and VPL nucleus (***B***), and spinothalamic tract below the thalamus (***C*** and ***D***).

(2) **Slow pain fibers** (**C fibers**) cause dull, intense, aching, diffuse pain. The **neurotransmitter** for this pain is thought to be **substance P,** the release of which is **inhibited by opioids.**

c. **Referred pain** is pain felt in a structure away from the original irritation producing the pain.

(1) This type of pain follows the **dermatome rule,** which divides the body into segments innervated by nerves that arise from the same embryonic portion of the spinal cord.

(2) The most well-known example is **referral of cardiac pain to the inner aspect of the left arm.**

4. Mapping of the somatosensory cortex has identified **two somatic sensory areas: SI** and **SII.**
   a. **SI** is in the **postcentral gyrus,** and **SII** in the wall of the **sylvian fissure.**
   b. The arrangement of the thalamic fibers in SI is such that parts of the body have been mapped in order along the postcentral gyrus. This map of the body is called the **sensory homunculus,** in which the proportions of the homunculus have been distorted to correspond to the size of the cortical receiving areas (Figure 7–5).

LESIONS OF THE SOMATOSENSORY SYSTEM

- ***Lesions of the dorsal column–medial lemniscal system*** *result in loss of joint posture sensation, vibratory and pressure sensation, and touch discrimination.*
  –They are evaluated by testing vibratory sense.

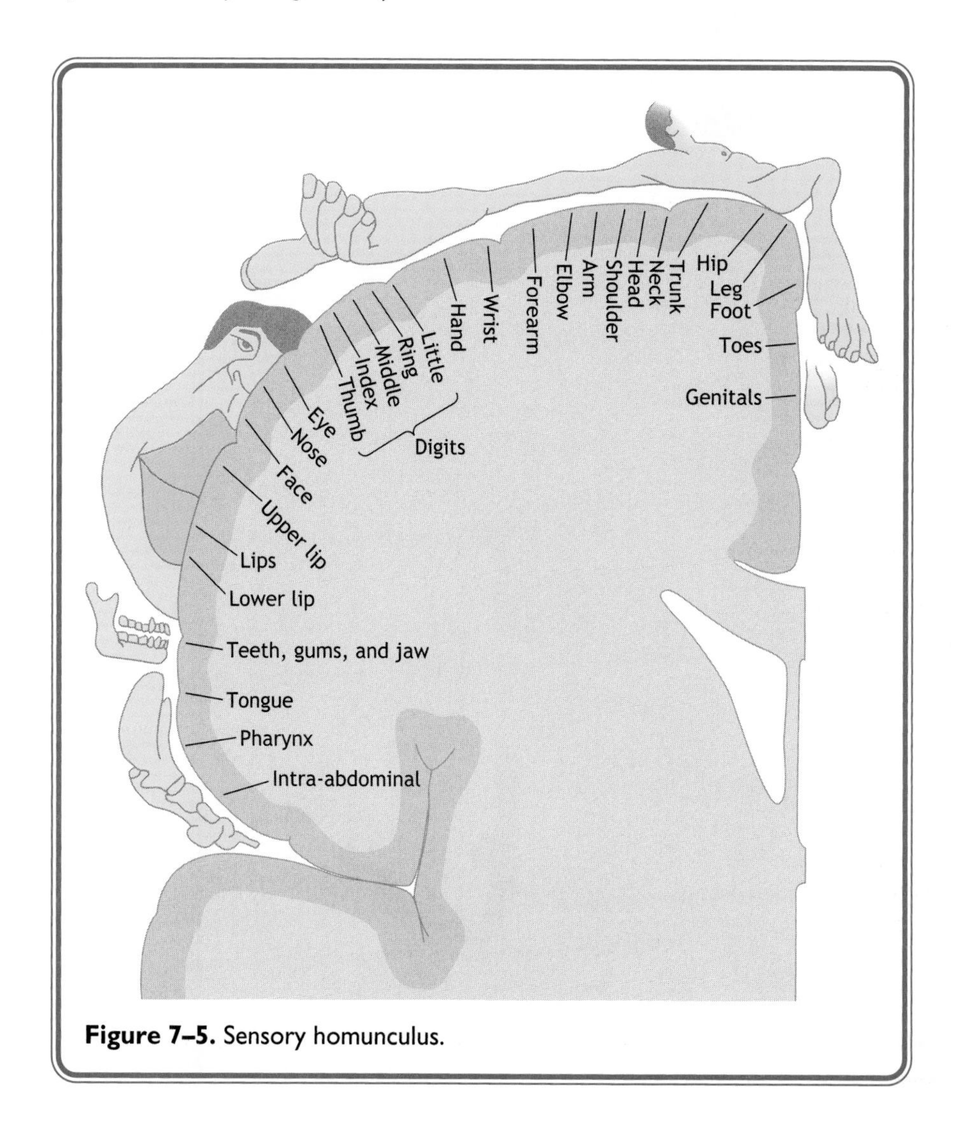

Figure 7–5. Sensory homunculus.

*–They are diagnosed by the **positive Romberg sign,** in which the patient sways when standing with feet together and eyes closed.*
*–Dorsal column lesions are **ipsilateral.***

- ***Lesions of the spinothalamic tract** result in contralateral loss of pain and temperature sensations.*
  –The patient experiences analgesia on one side, below the lesion location.
  –Thus, the lesion is on the contralateral side of the spinal cord or brainstem.

C. Visual Pathways (Figure 7–6)

1. Vision is based on principles of **optics.**
   a. The **major site of refraction** is the **anterior surface of the cornea.**

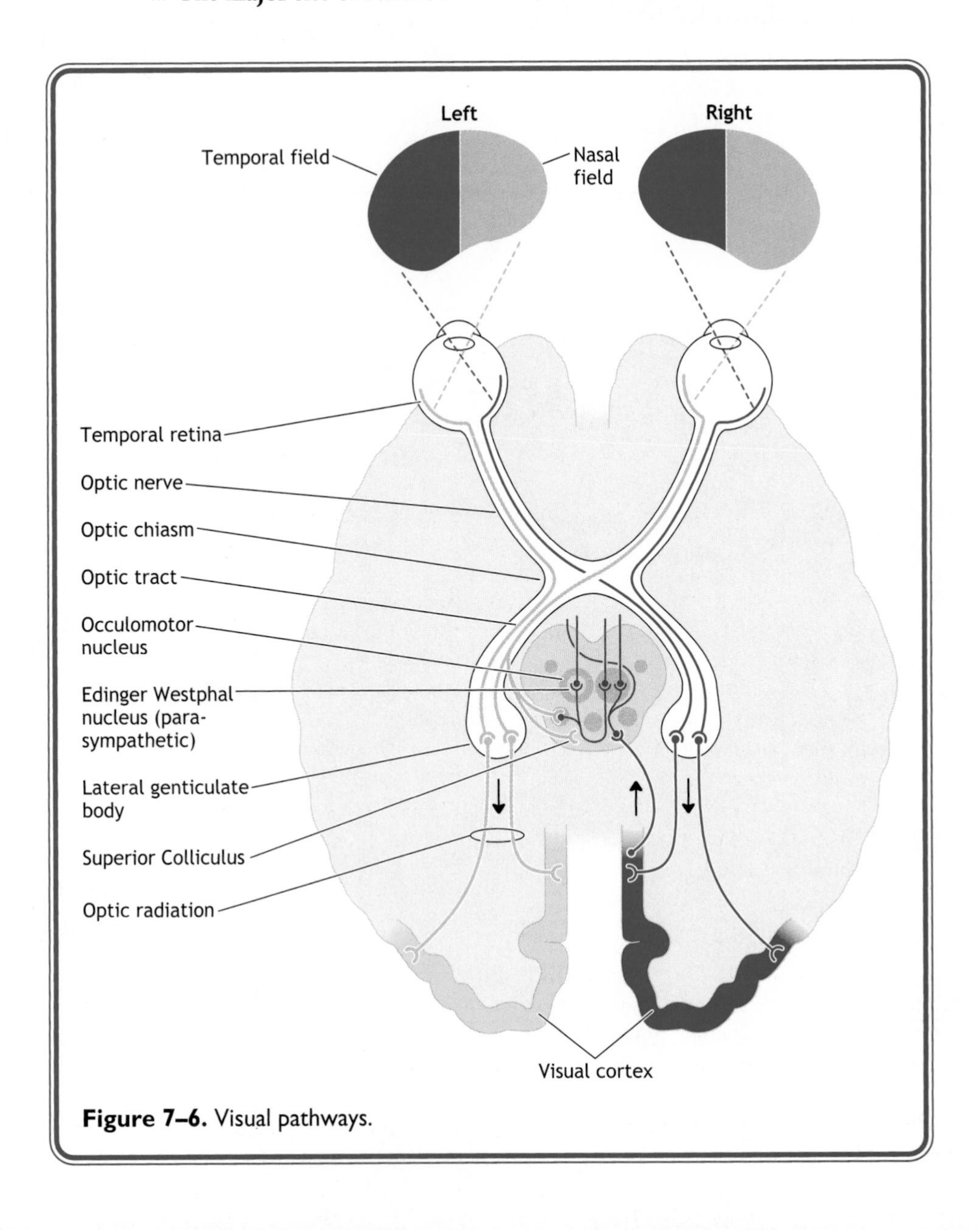

Figure 7–6. Visual pathways.

b. Light must pass through the **cornea, aqueous humor, pupil, lens,** and **vitreous humor** to reach the **retina.**
c. **Light** must then pass through the layers of the retina to reach the **photoreceptive layer of rods and cones.**

2. Vision difficulties are usually the result of **refractive problems.**
   a. **Emmetropia** is normal vision, in which light focuses on the retina.
   b. **Hyperopia** is farsightedness, in which light focuses behind the retina because the eyeball is too short. This problem is **corrected with convex lenses.**
   c. **Myopia** is nearsightedness, in which light focuses on the front of the retina because the eyeball is too long. This problem is **corrected with biconcave lenses.**
   d. **Astigmatism** refers to a nonuniform curvature of the lens, which is **corrected by a cylindrical lens.**
   e. **Presbyopia** is the reduced ability for accommodation that occurs with aging. It is **corrected with convex lenses.**
   f. The **myopic individual loses visual activity in the dark** because the pupil reflexly dilates, thus decreasing the depth of focus. In myopic individuals, the depth of focus depends on a small pupillary opening.
3. The retina comprises several layers and contains rods and cones plus neurons.
   a. The outer layers of rods and cones change light energy from **photons** into **membrane potentials.**
   b. Photopigments in rods and cones absorb photons, changing their molecular structure and reducing the amount of neurotransmitter released.
   c. Hence, **rods and cones release less neurotransmitter in the light** and **more in the dark.**
   d. **Rods and cones** have synaptic contacts on bipolar cells that **project to ganglion cells.**
   e. **Axons from ganglion cells** converge at the optic disc to **form the optic nerve,** which enters the cranial cavity through the optic foramen ending in the **lateral geniculate body of the thalamus.**
   f. **Optic tract fibers** also **project to the superior colliculi** for reflex gazes, **the pretectal area** for the light reflex, and **the suprachiasmatic nucleus** of the anterior hypothalamus for **generation of circadian rhythms.**
   g. At the **optic chiasm, most optic nerve fibers from the nasal half of each retina cross and project to the contralateral optic tract.**
   h. **Fibers from the temporal retina do not cross** at the chiasm and pass into the ipsilateral optic tract.
   i. Because the eye inverts images like a camera, each nasal retina receives information from a **nasal hemifield.**

OPTIC LESIONS (FIGURE 7–7)

- ***Cutting the optic nerve*** *results in* ***blindness in the ipsilateral eye.***
- ***An aneurysm of the right internal carotid artery*** *results in* ***right nasal hemianopsia.***
- ***Compression of the optic chiasm*** *by a pituitary tumor results in loss of peripheral vision in both temporal fields, or* ***bitemporal heteronymous hemianopsia.***
- ***Lesions of the optic tract*** *result in a visual loss from the contralateral visual field, or* ***homonymous hemianopsia.***
- ***Lesions inside the primary visual cortex*** *(geniculocalcarine tract) result in* ***contralateral homonymous hemianopsia,*** *sparing macular (central) vision.*

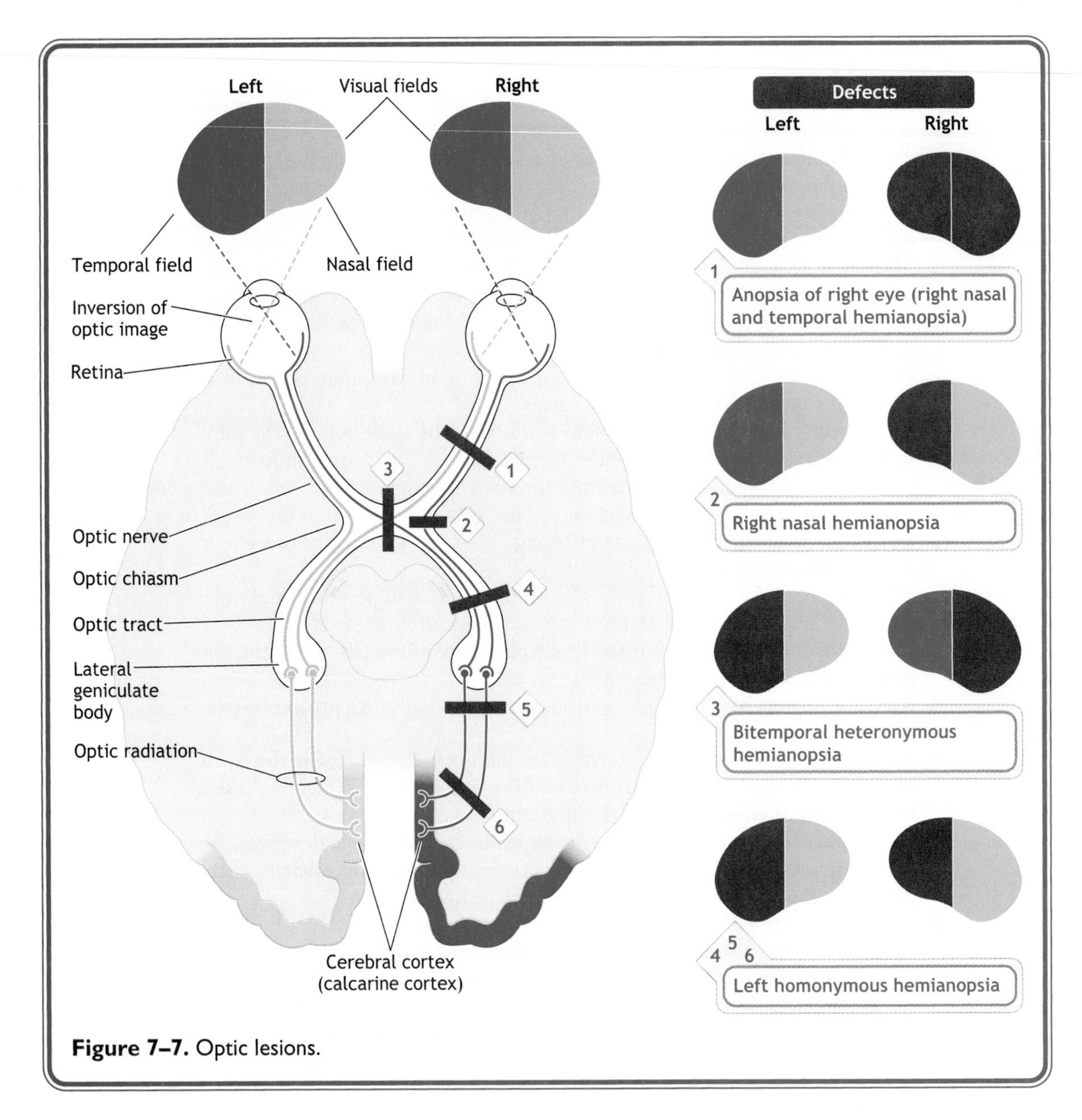

Figure 7–7. Optic lesions.

D. Auditory System (Figure 7–8)

1. The **external ear** includes the **pinna and external auditory meatus** and **directs sound waves to the tympanic membrane** (**eardrum**), causing it to vibrate.
2. The **middle ear** lies in temporal bone and contains the **auditory ossicles** (**malleus, incus, and stapes**).
   a. The **malleus** inserts into the **tympanic membrane;** and the **stapes** inserts into the **oval window,** a membrane **between the middle ear and inner ear.**

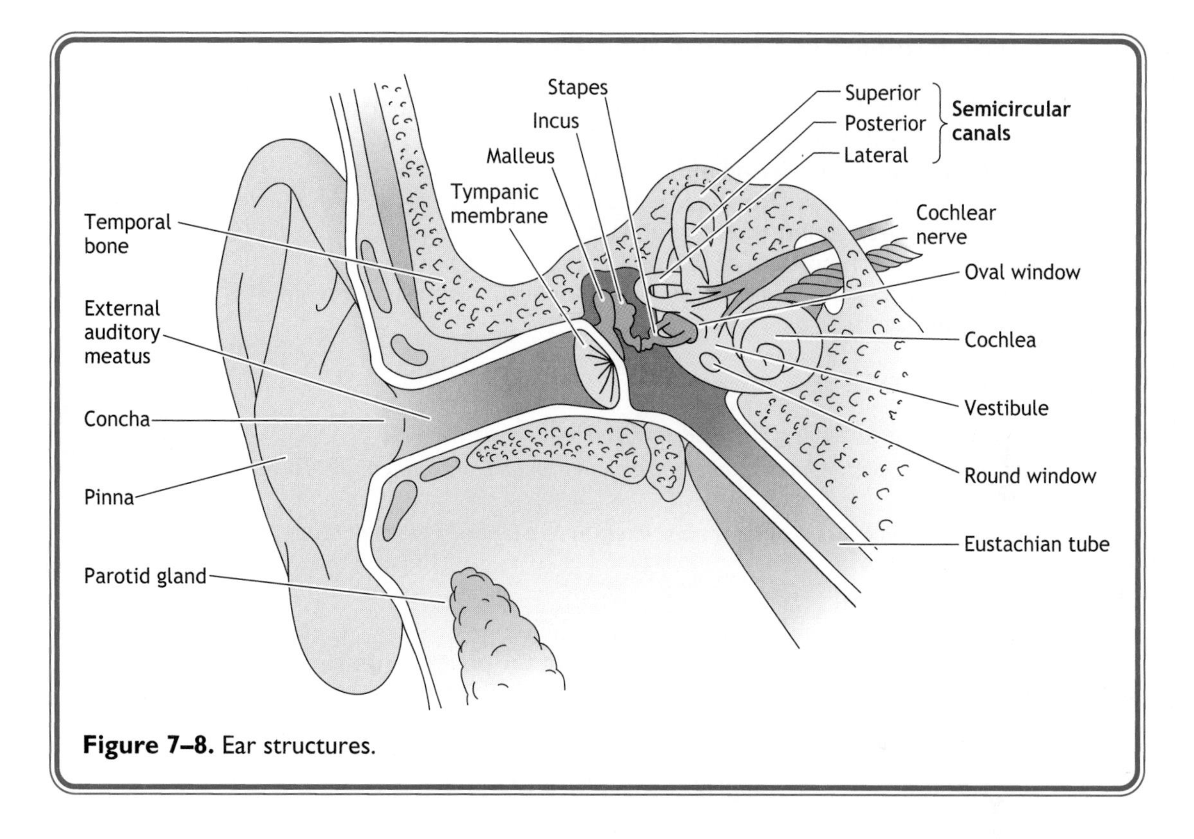

Figure 7–8. Ear structures.

b. Movement of the eardrum causes **vibrations of the ossicles** that are **transferred through the oval window to** the fluid of **the inner ear.**

c. The **middle ear cavity** communicates with the nasopharynx by the **eustachian tube,** allowing air pressure to be equalized on both sides of the tympanic membrane.

3. The **inner ear** consists of a labyrinth of interconnected sacs (the **utricle and saccule**) and channels (**semicircular canals** and the **cochlear duct**).
   a. The **cochlear duct, sacs, and semicircular canals** of the vestibular labyrinth are **filled with endolymph,** which is important for **hair cell function.**
   b. The **fluid outside** the vestibular labyrinth **is called perilymph.**
   c. The **cochlear duct** is the **auditory receptor of the inner ear.** It contains hair cells that respond to vibrations transmitted by the ossicles to the oval window.
   d. The **organ of Corti** is located in the **basilar membrane** and contains inner and outer hair cells with their cilia projecting into the **tectorial membrane.**
   e. The **spiral ganglion** contains cell bodies of the **auditory nerve** whose peripheral axons innervate hair cells on the **organ of Corti.**
   f. **Auditory pathway fibers ascend through the lateral lemniscus** to the inferior colliculus to the medial geniculate body and then to the auditory cortex.
4. **Auditory transduction** is the process of transforming sounds into nerve impulses.
   a. The **cilia** of hair cells **on the basilar membrane** are **embedded in the tectoral membrane.**

b. Sound waves produce **vibrations of the organ of Corti,** causing the hair cells to bend.

c. Bending of the hair cell stereocilia cilia causes cation channels to open or close whether depolarization or hyperpolarization occur. This results in an **oscillating cochlear microphonic potential.**

(1) Inner hair cells transduce sound while outer hair cells sharpen sound tuning

d. **High-frequency sound waves** cause maximum displacement of the basilar membrane and stimulation of hair cells at the **base of the cochlea.**

e. **Low-frequency sound waves** maximally stimulate hair cells at the **apex of the cochlea.**

LESIONS AND HEARING LOSS

- ***Lesions of the cochlea*** *of one ear result in profound unilateral hearing loss.*
- ***Lesions of other auditory structures*** *in the brainstem, thalamus, or auditory cortex result in bilateral hearing loss.*
- *Patients complaining of* ***hearing loss in one ear most likely have a lesion in the middle ear, inner ear, CN VIII, or cochlear nuclei.***

E. Chemical Senses: Smell

1. **Smell is a primitive sense** not well developed in humans.
2. Within each nostril, **receptor cells** are located **in the olfactory mucous membrane** (Figure 7–9).
3. To be detected, substances must be slightly water soluble and somewhat lipophilic.
4. The **receptor proteins are coupled to cytoplasmic G-proteins,** allowing the olfactory system to detect substances at a concentration of only one part per billion.

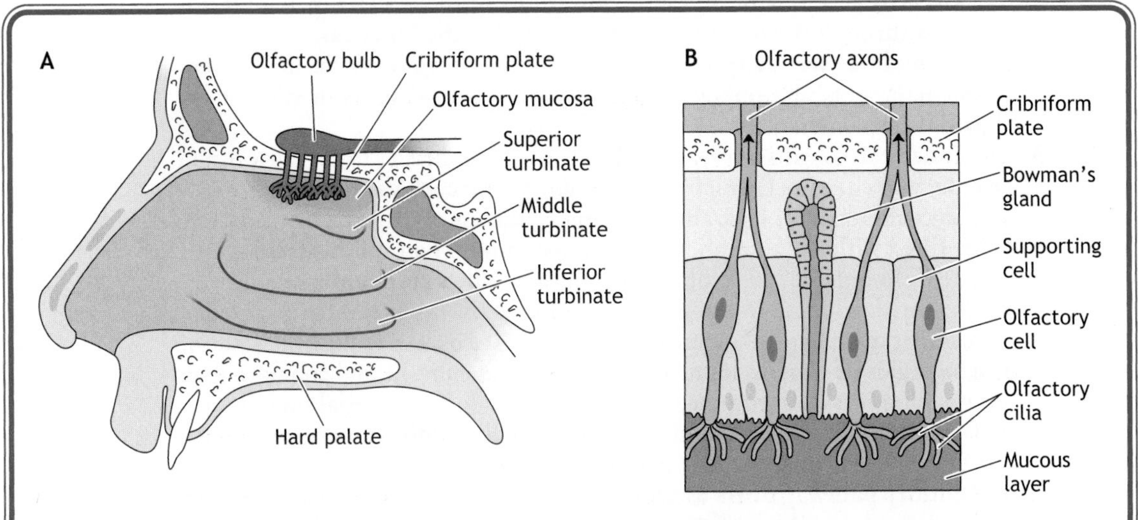

Figure 7–9. Olfactory system. The sensory organ for the sense of smell is the olfactory epithelium. It consists of ciliated receptor cells and supporting cells and is covered by a layer of mucus secreted by Bowman's glands, which lie beneath the epithelial layer. Axons from the olfactory receptor cells pass through the cribriform plate to synapse in the glomeruli of the olfactory bulbs.

5. The **G-proteins** activate adenylate cyclase and the production of cAMP, which opens Na^+ channels, producing **a depolarizing receptor potential.**
6. The receptor potential depolarizes the initial segment of the axon, propagating action potentials.
7. Receptor cells are true neurons that conduct action potentials to the CNS and are continuously replaced by basal stem cells of the olfactory epithelium.
8. The **axons of the olfactory receptor neurons penetrate the cribriform plate** and enter the olfactory bulb.
9. The three olfactory pathways that lead to the CNS are
   a. The pathway to the **medial olfactory area** anterior to the hypothalamus
   b. The pathway to the **lateral olfactory area** of the **pyriform cortex,** which is the **area of aversion development** (eg, smells inducing nausea)
   c. The pathway to the posterior part of the **orbitofrontal cortex,** which is an important **area for analysis of smells**
10. The olfactory epithelium is innervated by the **trigeminal nerve** (**CN V**), which detects noxious stimuli (eg, menthol).
11. Because olfactory nerves pass through the cribriform plate on their way to the olfactory bulb, **damage to the cribriform plate** can lead to a **loss of the sense of smell** (**anosmia**).
12. Olfactory **thresholds increase with aging,** resulting in a majority of 80-year-olds not being able to identify smells.

ANOSMIA

- *Anosmia is a loss of the sense of smell.*
- *Possible causes include* ***viral disease; trauma*** *(eg, fracture of the cribriform plate);* ***congenital hypogonadism*** *due to* ***Kallmann syndrome,*** *a gonadotropin-releasing hormone deficiency;* ***smoking;*** *or* ***tumor*** *(eg, olfactory meningoma).*

F. Chemical Senses: Taste

1. **Taste** occurs via sensory receptors primarily on the dorsal surface of the tongue known as **taste buds** (Figure 7–10).
2. Five primary sensations of taste are **sour, salt, sweet, bitter, and amino acids.**
3. Taste receptor cells are covered with microvilli, or taste hairs, that project into the **taste pore** that opens on the tongue surface.
4. Taste buds are found in the walls of the **fungiform, folate,** and **vallate** papillae on the tongue, epiglottis, palate, and pharynx. The **filiform papillae** on the back of the tongue do not contain taste buds.
5. **Fungiform papillae** are located on the anterior two-thirds of the tongue and detect **sweet** and **salty sensations** and are innervated by the **facial,** or **chorda tympani, nerve** (**CN VII**).
6. **Vallate** and **folate papillae** are located on the posterior one-third of the tongue and are innervated by the **glossopharyngeal nerve** (**CN IX**).
7. Binding of a taste substance to the specific receptor proteins or the taste hairs opens Na^+ channels, creating a receptor potential for that **taste,** which is **transmitted via the facial, glossopharyngeal, or vagus nerves.**
8. The **taste fibers** unite and **ascend the tractus solitarius** in the medulla, where they **synapse on second-order neurons.**
9. **They project to the ventral posteromedial nucleus** of the thalamus **and are then relayed to** the taste projection area of **the cerebral cortex.**

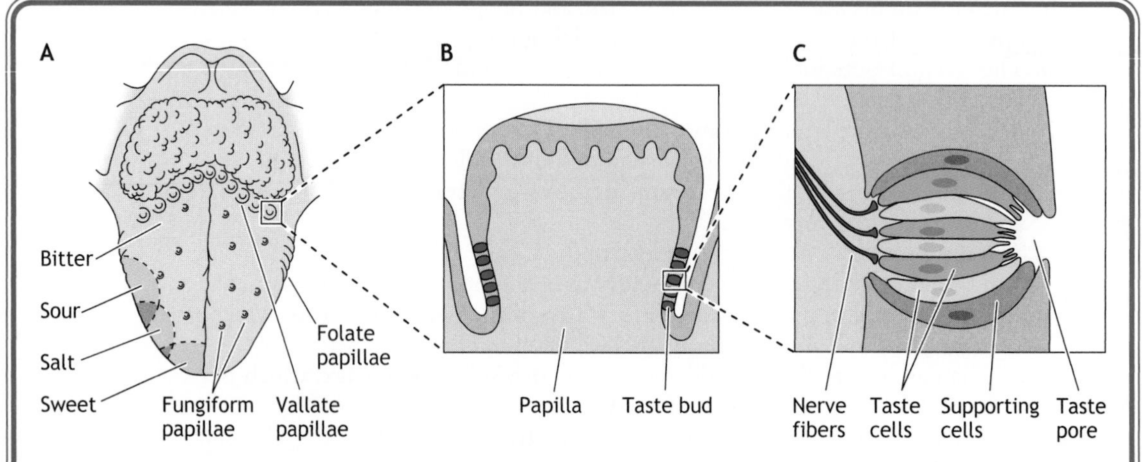

Figure 7–10. Taste system. ***A.*** The tongue is the primary organ of taste and is covered by surface papillae (filiform, folate, fungiform, and vallate). Taste buds are found on the folate, fungiform, and vallate papillae only. ***B.*** Taste buds are located on the side below the surface epithelium and consist of taste cells and supporting cells. ***C.*** Taste buds open to the outside by taste pores, and taste cells are innervated by facial and glossopharyngeal nerve fibers.

10. Although taste cells respond to more than one taste, the response is stronger for one modality than others.
11. Taste plays an **important role in food selection and the regulation of some GI secretions.**

G. Vestibular System

1. The **vestibular system** contains two kinds of sensory receptors:
   a. **Hair cells in a macula** (otolithic organ) are found in the **utricle** and **saccule. Each macula responds to linear acceleration** and detects head positional changes relative to gravity.
   b. **Hair cells** in the three **semicircular canals detect changes in angular acceleration** resulting from circular movements of the head.
2. Four **vestibular nuclei** are located in the rostral medulla and caudal pons.
   a. The vestibular nuclei receive afferents from the vestibular nerve that innervates receptors located in the semicircular canals, utricle, and saccule.
   b. The vestibular nuclei have two efferent tracts: the **vestibulospinal tract** and the **medial longitudinal fasciculus.**
3. **Primary vestibular fibers terminate in** the vestibular nuclei and **the flocculonodular lobe of the cerebellum.**
4. **Secondary vestibular fibers** from the vestibular nuclei **supply** the **motor nuclei** of CNs III, IV, and VI **and are involved in** the production of **eye movements.**
5. The **vestibuloocular reflex** is the slow movement of the eyes in a direction opposite to the direction of the rotation of the body **due to stimulation of the horizontal semicircular canals.**
   a. The reflex keeps the eyes fixed on a stationary point as the head rotates and involves several steps (Figure 7–11):
      (1) If the **head is rotated** horizontally **to the left,** ***hair cells in the left semicircular canals are stimulated.***

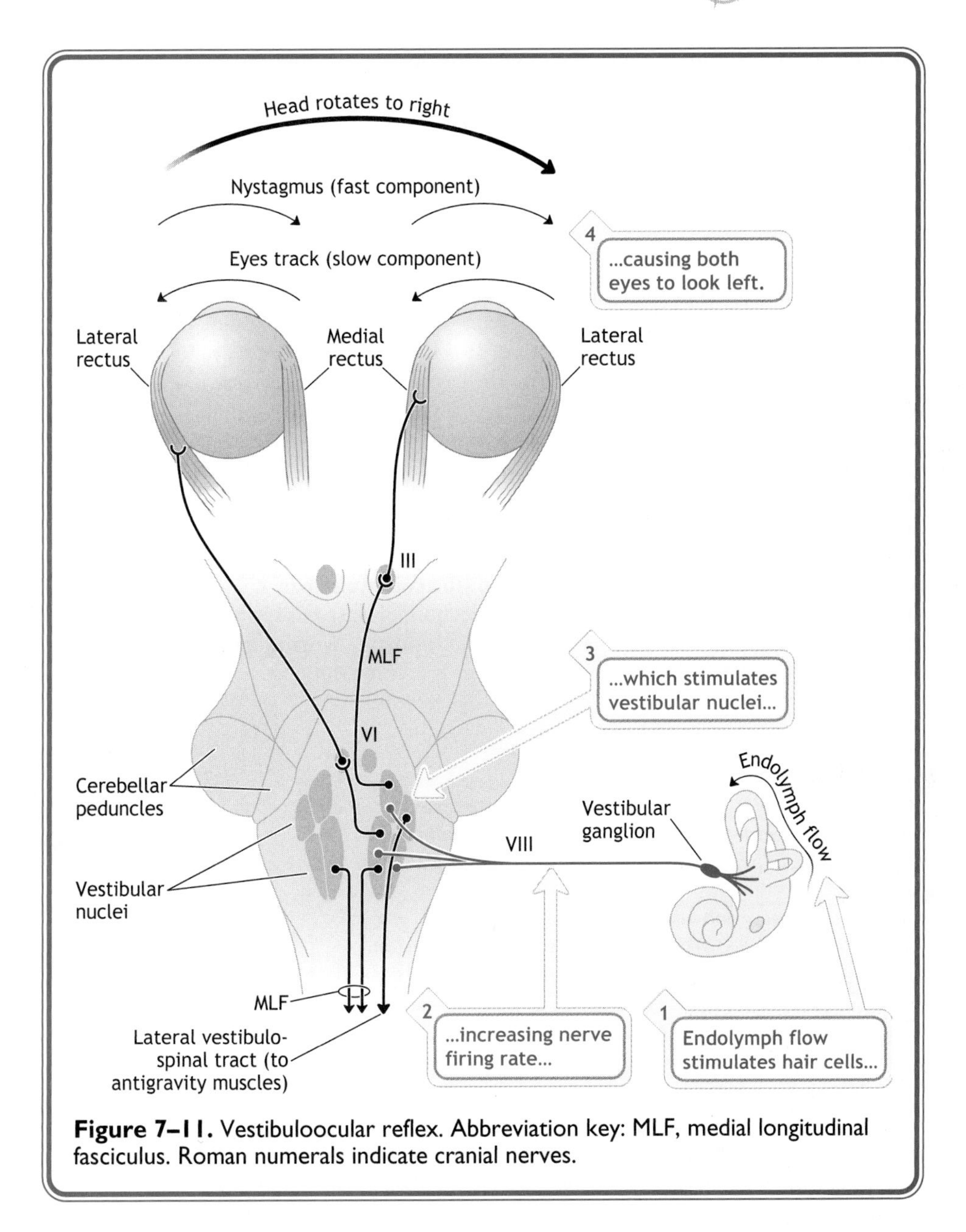

Figure 7–11. Vestibuloocular reflex. Abbreviation key: MLF, medial longitudinal fasciculus. Roman numerals indicate cranial nerves.

(2) The **left eighth nerve** increases its firing rate to the **left vestibular nuclei.**

(3) The **left oculomotor nerve** to the left medial rectus muscle adducts the left eye, and the **right abducens nerve** to the right lateral rectus muscle abducts the right eye, resulting in **both eyes looking to the right.**

b. The rhythmic oscillation of the eye at the start and end of a period of rotation is called **nystagmus.** Nystagmus is defined by the direction of the rapid reflex movement or the **fast component.**

(*1*) **Postrotatory nystagmus** is rapid eye movement that develops in the direction opposite to the previous head rotation when rotation is stopped.

(*2*) **Nystagmus** is seen clinically in patients with brainstem lesions. In **pathologic vestibular nystagmus,** the **slow component** (movement is to the side of the lesion) is seen without head movement. The cortex responds by moving both eyes quickly (**fast component**) back in the opposite direction.

c. Irrigation of the ear with water that is above or below body temperature causes convection currents in the endolymph. This **caloric stimulation test** is used to test the integrity of the vestibuloocular reflex in comatose patients and produces **nystagmus, nausea,** and **vertigo.** It is important, therefore, to use fluid at body temperature when irrigating the ear to treat ear infections.

(*1*) **Warm water** normally stimulates the horizontal semicircular canal, causing the eyes to move slowly in the opposite direction away from the warm water ear. The **fast component moves the eyes** quickly back **toward the warm water ear.**

(*2*) **Cool water** normally inhibits the horizontal semicircular canal, causing the eyes to move slowly toward the cool water ear. The corrective (**fast component**) of nystagmus **moves the eyes away from the cool water ear.**

(*3*) The mnemonic **COWS** summarizes the direction of **fast-component vestibular nystagmus** in the caloric test: **Cool, Opposite, Warm, Same.**

(*4*) In coma, the brainstem is depressed and there is a complete absence of movement.

VERTIGO

- *Vertigo is the sensation that the **environment is spinning while the eyes are open.** By contrast, **dizziness is the sensation that the individual,** not the surroundings, **is spinning.***
  *–Vertigo may result from **peripheral** (end organ or nerve) or **central** (nuclear or brainstem pathway) **vestibular structure lesions.***
  *–Vertigo is **severe in peripheral** disease and **mild in central** disease.*
  *–**Chronic vertigo** (ie, more than 2–3 weeks in duration) **suggests a central lesion.***
- ***Ménière disease** involves abrupt recurrent attacks of vertigo that last minutes to hours.*
  *–It is associated with **hearing loss** and **tinnitus** (ringing in ears).*
  *–It usually involves **only one ear** and may result from **endolymph overproduction.***
  –Treatment is surgical (ie, vestibular neurectomy) or pharmacologic (eg, anticholinergics, antihistaminergics, diuretics, barbiturates, diazepam).
- ***Benign paroxysmal positional vertigo** is a sudden episode of vertigo associated with head movement. The vertigo lasts for seconds and may be related to otolith displacement in the inner ear.*
- ***Motion sickness** is due to a **conflict between** information from the **vestibular** system **and other sensory systems.***
  *–It is **not due to vestibular system damage.***
  *–**Patients with bilateral damage to the vestibular system do not exhibit motion sickness.***

III. Motor Pathways

A. Organization

1. An **upper motoneuron** and a **lower motoneuron** form the basic neural circuit in the voluntary movement of skeletal muscle.

2. Motor pathways start from regions in the cerebral cortex where a group of neurons control the contraction of individual muscles (ie, **motor cortex**).

3. The motor cortex makes up the posterior third of the frontal lobes and is responsible for generating movement.
4. Excited **upper motoneurons** in the **primary motor cortex** direct lower centers such as the **basal ganglia, cerebellum,** and **brainstem** to make specific, often preprogrammed responses.
5. Anterior to the primary motor cortex are the **premotor and supplementary motor cortical areas,** which set the stage for movement executed by the primary motor cortex.
6. The premotor cortex contains specialized areas involved in specific motor functions, such as **Broca's area,** which controls word formation.
7. The motor cortex receives input from the **somatic sensory cortex** and from auditory and visual pathways to initiate appropriate motor responses.
8. **The corticospinal (pyramidal) tract** is the primary efferent path from the cortex.
9. **Upper motoneurons** are those of the **corticospinal tract from the primary motor cortex down to the spinal cord.**
10. The **corticospinal tract descends as the lateral corticospinal tract** in the spinal cord. As it descends, **axons** leave the tract and **enter the gray matter of the ventral horn to synapse on lower motoneurons.**
11. **Lower motoneurons** are those that **travel from the anterior horn of the spinal cord to innervate specific muscles.**
12. Thus, to initiate a voluntary contraction of skeletal muscle, the upper motoneuron innervates the lower motoneuron, and the lower motoneuron innervates the skeletal muscle.

UPPER VERSUS LOWER MOTONEURON LESIONS

- ***Upper Motoneuron Lesions***
  –Clinical findings include spastic paralysis, hyperreflexia, increased muscle tone, muscle weakness, disuse atrophy of muscles, and decreased speed of voluntary movements.
  *–The **Babinski sign** is present, characterized by extension of the great toe and fanning of other toes when the sole of the foot is stroked.*
  –These lesions affect a large area.
- ***Lower Motoneuron Lesions***
  –Clinical findings include flaccid paralysis, absence of reflexes (areflexia), decreased muscle tone, muscle atrophy, and loss of voluntary movements.
  –The Babinski sign is not present.
  –These lesions affect a small area.

B. Motor Units

1. A **motor unit** is made up of a single motoneuron and the muscle fibers it innervates.
2. A **motoneuron pool** is the group of motoneurons innervating fibers in the same muscle.
3. The **force** of muscle contraction **depends on** the tension generated and the **number of motor units recruited.**
4. The larger the motoneuron, the greater the number of muscle fibers innervated and the larger the force generated.

C. Muscle Fibers

1. **Extrafusal fibers** innervated by α **motoneurons (large cells in the ventral horn)** are the most plentiful and provide the force for muscle contraction.

2. **Intrafusal fibers** innervated by γ **motoneurons** are encapsulated in sheaths to **form muscle spindles** and are too small to generate force for muscle contraction.
   a. **Nuclease bag fibers** are intrafusal fibers innervated by **group Ia afferents** that detect the **rate of change in muscle length** and have nuclei concentrated in a central bag-like region.
   b. **Nuclear chain fibers** are intrafusal fibers innervated by **group II afferents** that detect **static changes in muscle length** and have nuclei arranged in rows.
   c. γ **motoneurons** adjust the sensitivity of the muscle spindle to provide appropriate responses during muscle contraction.

D. Function of Muscle Spindles (Figure 7–12)

1. **Muscle spindles** act as sensory receptors in skeletal muscle stretch reflexes.
   a. They detect both **static and dynamic changes** in muscle length.
   b. The **finer the movement** required, the **more muscle spindles** a muscle contains.
2. Muscle spindle reflexes oppose increases in muscle stretching.
   a. When muscle length is increased (stretched), the muscle spindle is stretched, stimulating **afferent groups Ia and II.**
   b. **Group Ia afferents stimulate** α **motoneurons** in the **spinal cord,** causing **muscle contraction** and shortening of the muscle.
3. Both ends of the muscle spindle are connected in parallel with extrafusal fibers so that their length and rate of change in length can be monitored.

E. Muscle Reflexes

1. The **muscle stretch (myotatic) reflex** (Figure 7–13A) is a **stereotyped muscle contraction** in response to a stretch of that muscle.
   a. This reflex is the primary mechanism for regulating muscle tone (tension is present in all resting muscle).
   b. Stretching of muscle spindles activates group Ia afferents that synapse with α motoneurons in the spinal cord, causing muscle contraction.

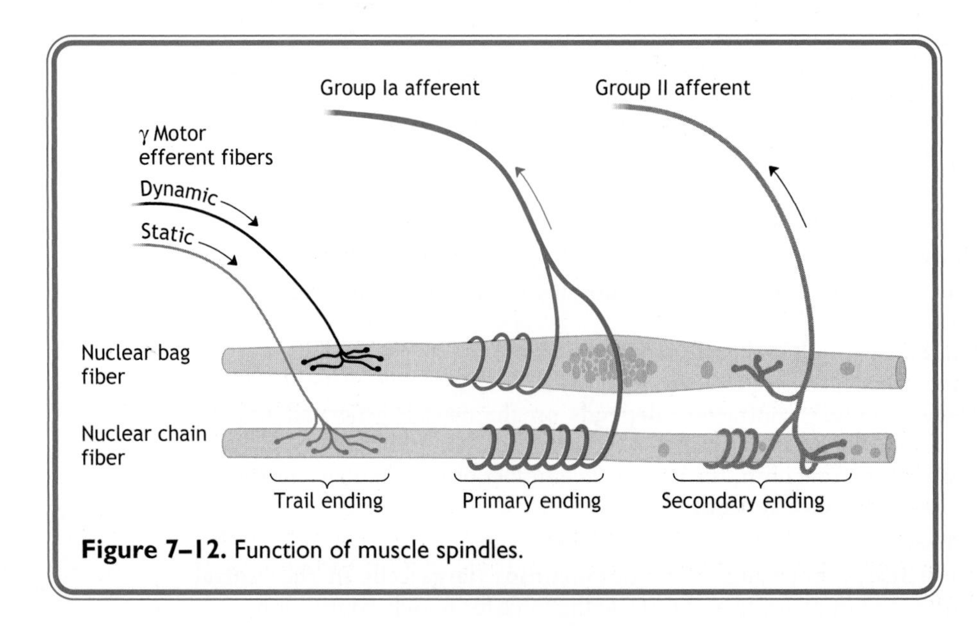

Figure 7–12. Function of muscle spindles.

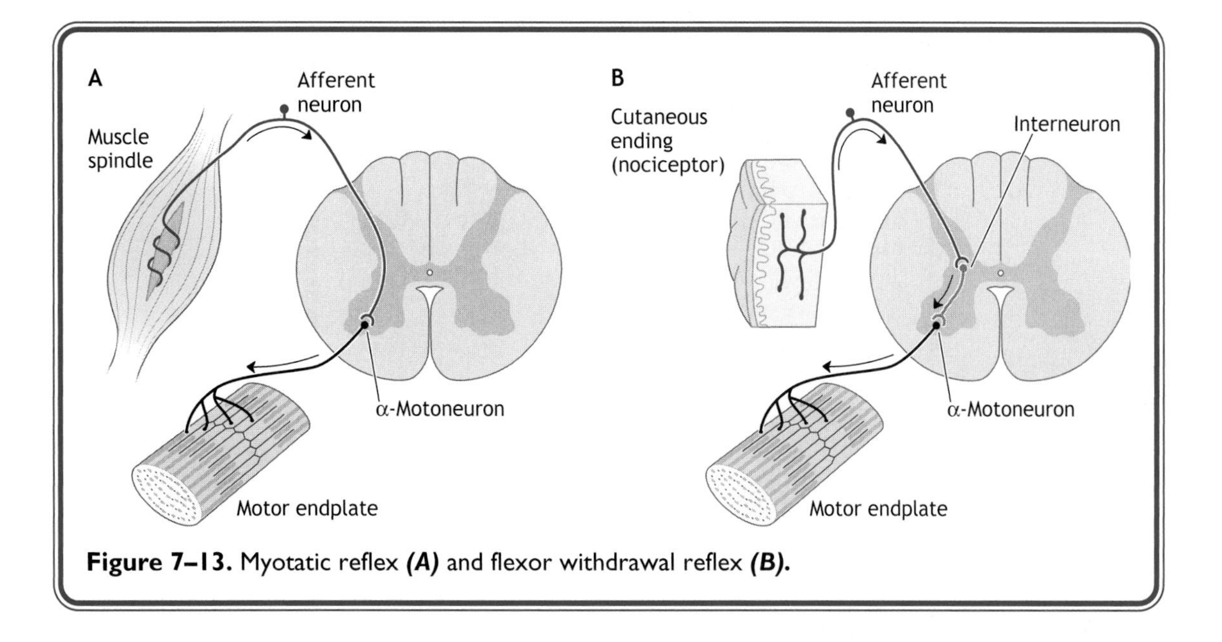

Figure 7–13. Myotatic reflex ***(A)*** and flexor withdrawal reflex ***(B)***.

c. The best example is the **knee-jerk reflex** stimulated by tapping the patellar ligament that stretches the quadriceps muscle, causing a sudden extension of the leg at the knee.

2. The **flexor withdrawal reflex** (Figure 7–13B) is a **protective reflex** in which a usually **painful stimulus causes withdrawal of a stimulated limb.**
   a. This reflex **may be accompanied by a crossed extension reflex,** in which the contralateral limb is extended to support the body.
   b. Flexor reflex **afferent groups II, III, and IV** synapse polysynaptically onto motoneurons in the spinal cord.
   c. Because of the persistent neural activity in the polysynaptic circuits, an **afterdischarge** occurs that prevents muscle relaxation.
3. The **inverse muscle stretch reflex** (Figure 7–14) is associated with Golgi tendon organs arranged in series with extrafusal muscle fibers that **detect muscle tension.**
   a. **Golgi tendon organs** respond to increases in force or tension generated by muscle contraction that **increases** the firing rate of **group Ib afferent neurons.**
   b. **Group Ib afferent neurons** that innervate the Golgi tendon organs polysynaptically also **facilitate antagonists** and **inhibit agonist muscles.**
   c. Muscle tone and reflex activity are influenced by γ **motoneurons** that **directly innervate muscle spindles** and regulate their sensitivity to stretch.
   d. **Upper motoneurons innervate γ motoneurons** and influence the sensitivity of muscle spindles to stretch.

F. **Spinal Cord Organization (Figure 7–15)]**

1. Inside the spinal cord, **gray matter** is centrally located in the shape of a butterfly and **contains neuron cell bodies,** their **dendrites,** and the **proximal parts of axons.**

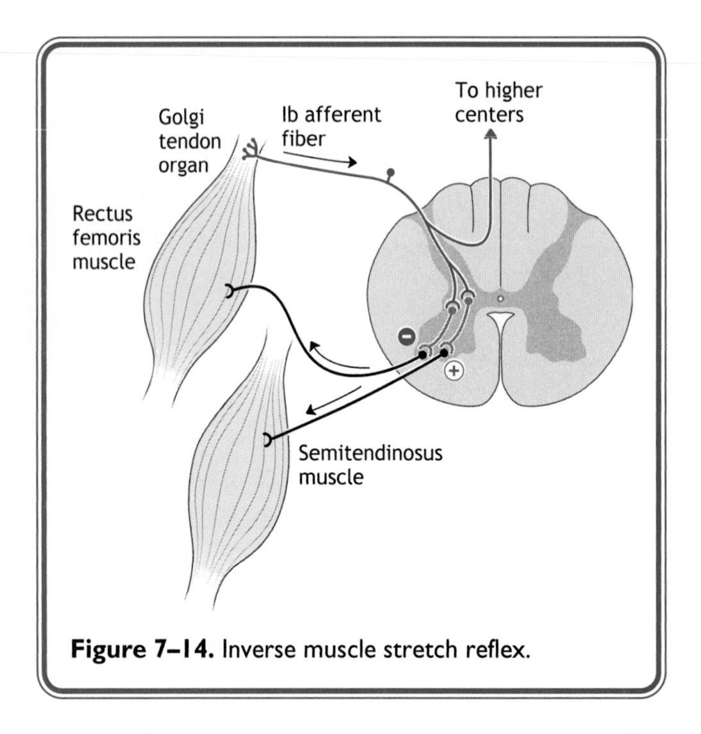

Figure 7–14. Inverse muscle stretch reflex.

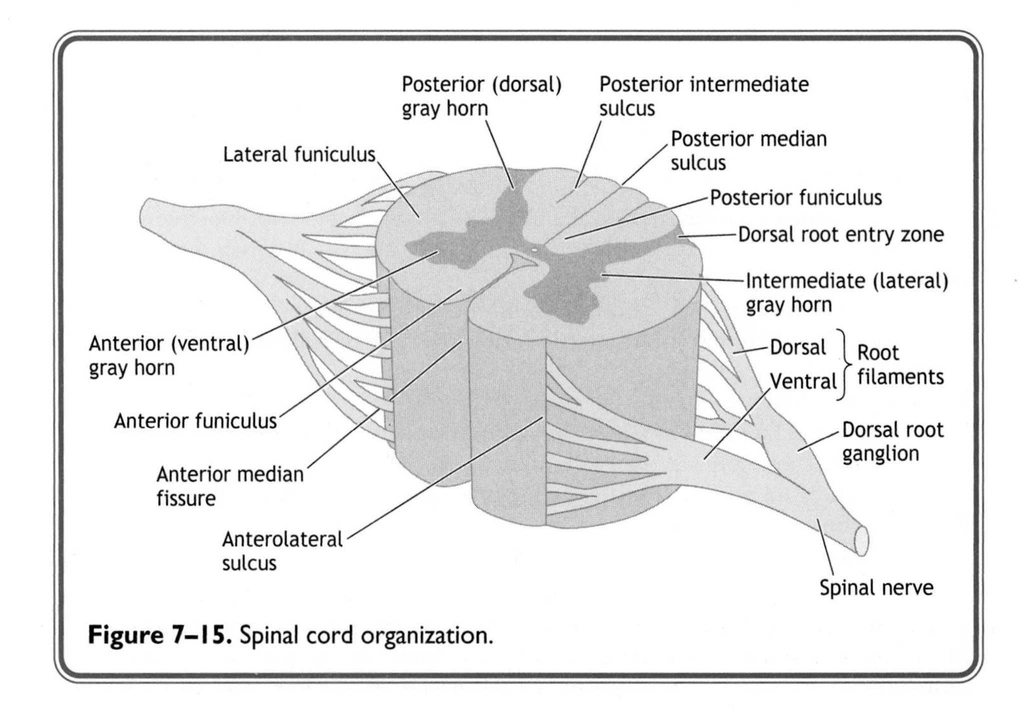

Figure 7–15. Spinal cord organization.

a. **White matter** surrounds the gray matter and **contains** bundles of functionally similar axons called **fasciculi** or **tracts,** which ascend or descend in the spinal cord.
b. The gray matter is organized into a **dorsal horn,** a **ventral horn,** and **an intermediate zone.**

2. The **dorsal horn** contains neurons that respond to sensory stimulation (eg, pain and temperature). **Dorsal horn neuron cell bodies project** to higher levels of the CNS (ie, **brainstem, cerebral cortex,** and **cerebellum**).
3. The **ventral horn** contains α and γ motoneurons.
   a. **α motoneurons** have axons that collect in bundles that **leave the ventral horn** and pass through the ventral white matter before entering the ventral root.
   b. **α motoneurons innervate skeletal muscle** (**extrafusal fibers**), and **γ motoneurons innervate intrafusal fibers** of the muscle spindle.
   c. Some axons send off branches that turn back into the spinal cord and synapse with small interneurons called **Renshaw cells.** When they are stimulated, **Renshaw cells inhibit the motoneuron** (**negative feedback**).

G. Effects of Spinal Cord Transection

1. **Spinal shock** is the loss of spinal reflexes immediately following injury to the spinal cord; it involves descending motor pathways.
   a. Spinal shock is probably **due to loss of normal excitatory input** from higher centers (ie, the vestibulospinal, reticulospinal, and corticospinal tracts).
   b. The interval between cord transection and the initial return of reflex activity is **about 2 weeks,** as excitability of undamaged neurons increases.
   c. Spinal shock involves loss of excitatory influence from α and γ motoneurons.
2. **Paraplegia** refers to loss of function of the legs and pelvic organs, and **quadriplegia** or **tetraplegia** refers to loss of motor and sensory function in the arms, trunk, legs, and pelvic organs.
   a. **Lesions at C3** cause breathing to stop because respiratory muscles have been cut off from brainstem control centers.
   b. **Lesions at C7** interrupt sympathetic tone to the heart, resulting in decreased heart rate and blood pressure.
   c. **Lesions below T12** result in flaccid paralysis of affected skeletal muscle groups and of the muscles controlling bowel, bladder, and reproductive function.

H. Transection Above the Spinal Cord

1. Lesions that isolate the hindbrain and the spinal cord from the rest of the brain (eg, **lesions above the lateral vestibular nuclei**) cause **decerebrate rigidity,** due to removal of inhibition by higher centers.
2. **Lesions above the red nucleus** result in a loss of the cortical area that inhibits γ efferent activity, causing **decorticate rigidity,** which is seen only at rest because it is otherwise obscured by phasic postural reflexes.

SELECTED SPINAL CORD LESIONS

- ***Brown-Séquard Syndrome***
  –The syndrome involves hemisection of the spinal cord due to a ***posterior white column lesion.***
  –It causes ***ipsilateral loss*** *of touch, tactile, and vibration sense below the lesion and* ***contralateral loss of pain and touch due to loss of the spinothalamic tract.***
  –Corticospinal tract lesions produce ***ipsilateral spastic paresis*** *(slight paralysis) below the lesion.*

*–Loss of lower motoneurons produces **ipsilateral flaccid paralysis** at the level of the lesion.*
*–If the **lesion occurs above T1, Horner syndrome** occurs, characterized by **ptosis** (drooping of upper eyelids), **myosis** (contraction of the pupil), and **anhydrosis** (deficiency in sweating) on the side of the lesion.*

- ***Subacute Combined Degeneration** (**Vitamin B_{12} Deficiency, Pernicious Anemia**)*
  *–Posterior white column lesions cause **bilateral loss of touch, vibration,** and **tactile sense.***
  *–Corticospinal tract lesions cause **bilateral spastic paresis** below the lesion.*
- ***Syringomyelia***
  *–Spinothalamic tract lesions produce **bilateral loss of pain and temperature** one level below the lesion.*
  *–**Bilateral flaccid paralysis** occurs at the level of the lesion due to loss of lower motoneurons.*
- ***Amyotrophic Lateral Sclerosis** (**ALS,** or **Lou Gehrig Disease**)*
  *–ALS involves **combined upper motoneuron and lower motoneuron lesions** of the corticospinal tract.*
  *–**Progressive spinal muscular atrophy** (**ventral horn**) occurs.*
  *–**Flaccid paralysis** occurs in the upper limbs, whereas **spastic paralysis** occurs in the lower limbs.*

I. The Cerebellum or "little brain"

1. The **cerebellum** lies caudal to the occipital lobe and is involved in the planning and fine-tuning of skeletal muscle contractions.
2. It can be divided into three major lobes that correlate with function by transverse fissures:
   a. The **anterior fissure** separates the **anterior and posterior lobes.**
   b. The **posterolateral fissure** separates the small **flocculonodular lobe** from the posterior lobe.
      (1) The **flocculonodular lobe,** or **vestibulocerebellum,** controls balance and eye movement.
      (2) Input is to the vestibular nuclei.
3. A functional separation consists of a midline zone (**vermis**), which separates the two lateral cerebellar hemispheres.
   a. The **vermis and intermediate zones,** or **spinocerebellum,** control rate, force, and direction of movement with principal input to the spinal cord.
   b. The **hemisphere,** or **pontocerebellum,** is involved in the planning and initiation of movements; principal input is to the cerebral cortex.
4. The cerebellar cortex has three layers:
   a. The **molecular layer** is the outer layer and contains **basket** and **stellate cells,** as well as **parallel fibers,** which are the axons of the granule cells. Dendrites of the **Purkinje cell** extend into this layer.
   b. The **Purkinje layer** is the middle and **most important layer** because all inputs to the cerebellum are directed toward influencing the firing of the Purkinje cells, and only Purkinje cell axons leave the cerebellar cortex. Output is always inhibitory.
   c. The **granular layer** is the innermost layer and comprises Golgi type II cells, granule cells, and **glomeruli.** Each glomerulus has a granule cell, which is the **only excitatory neuron** in the cerebellar cortex.
5. The cerebellar cortex has two major afferents:
   a. **Climbing fibers** originate from the inferior olivary nuclear complex on the contralateral side of the medulla.
      (1) Climbing fibers provide **excitatory input to Purkinje cells** by synapsing on them.
      (2) They play a role in cerebellar **motor learning.**

b. **Mossy fibers** arise from the spinal cord and brainstem and represent axons from all other sources of cerebellar input.
   (1) They provide an indirect, **diffuse excitatory input to Purkinje cells** as well as **inhibitory neurons** (ie, **stellate, basket,** and **Golgi type II cells**).
   (2) All mossy fibers exert an **excitatory effect** on **granule cells,** which give rise to parallel fibers that stimulate **Purkinje cells.**

6. The cerebellar cortex has four major efferents (Figure 7–16):
   a. **Purkinje cells** are the only output and are always inhibitory via the neurotransmitter **γ-aminobutyric acid** (**GABA**).
   b. The **spinocerebellum** has efferents to the **red nucleus** and **reticular formation,** which influences lower motoneurons, via the reticulospinal and rubrospinal tracts, to adjust posture and effect movement.
   c. The **pontocerebellum** has efferents first to the thalamus and then to the cortex and influences lower motoneurons via the corticospinal tract. These efferents produce precise, sequential voluntary movements.

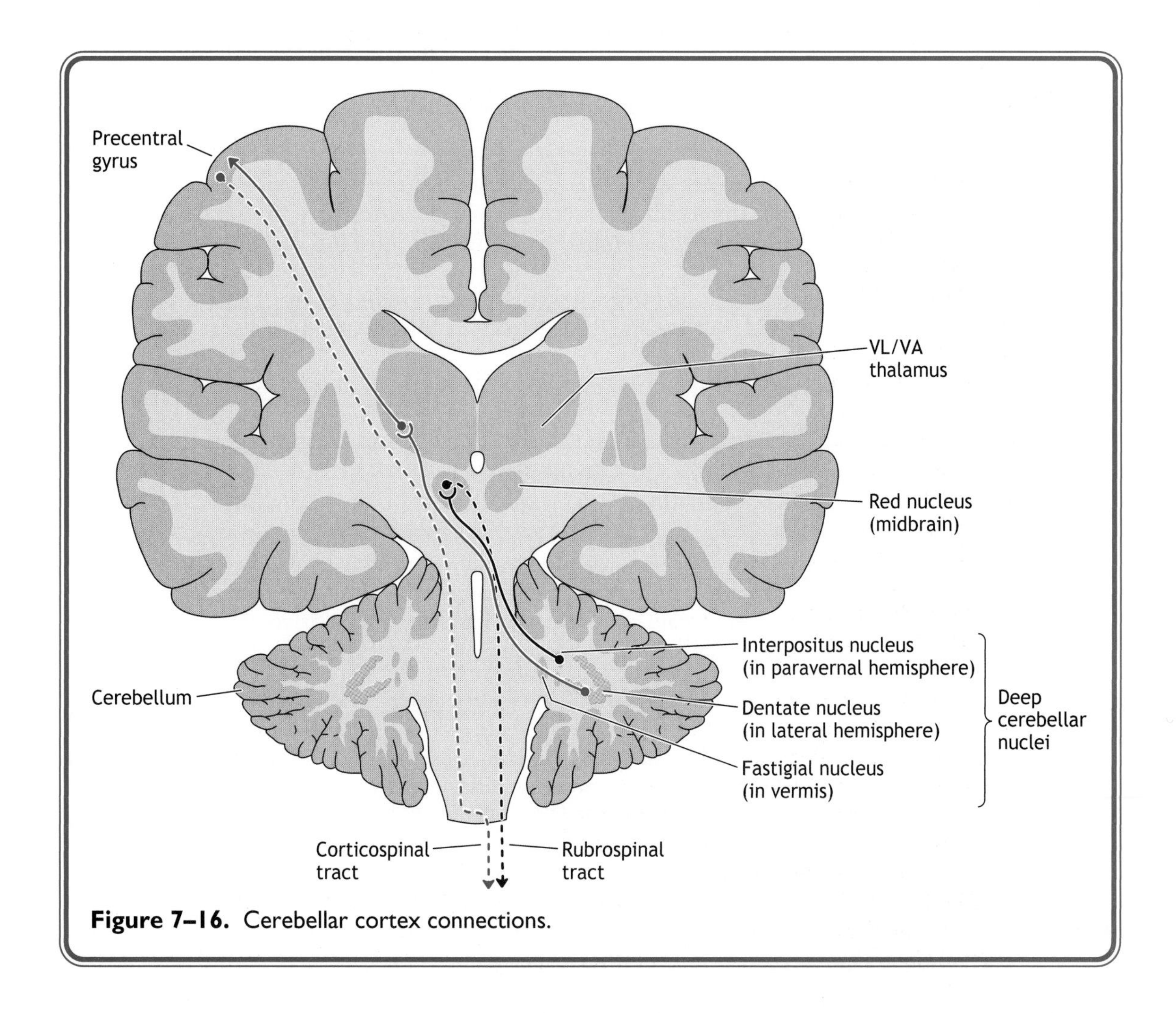

Figure 7–16. Cerebellar cortex connections.

d. The **vestibulocerebellum** has efferents to the vestibular nucleus and elicits positional changes of the eyes and trunk in response to head movements.

CEREBELLAR LESIONS

- ***Hallmark of cerebellar dysfunction is intention tremor*** *(a tremor with intended movement without paralysis or paresis); tremor is absent at rest.*
- ***Cerebellar lesions are expressed ipsilaterally.*** *Thus, patients with unilateral lesions fall toward the side of the lesion.*
- ***Lesions of the vermal region*** *result in difficulties maintaining posture, gait, or balance.*
  –These lesions are differentiated from dorsal column lesions by the ***Romberg sign.***
  –Patients with ***cerebellar lesions sway with their eyes open,*** *whereas those with* ***dorsal column lesions sway with their eyes closed*** *(****positive Romberg sign****).*
- ***Lesions that include the cerebellar hemisphere*** *produce a number of dysfunctions in addition to intention tremor.*
  *–****Dysmetria*** *is the inability to stop a movement at the proper place.*
  *–****Dysdiadochokinesia*** *is the decreased ability to perform rapid alternating movements.*
  *–****Scanning dysarthria*** *is asynchronous movement of muscles of speech, causing patients to divide words into syllables that disrupt the rhythm of speech.*
  *–****Gaze dysfunction*** *is oscillation of the eyes before fixing on the target point.*

J. Basal Ganglia

1. **Basal ganglia** initiate and control skeletal muscle movement.
2. Components of the basal ganglia include the **striatum,** which consists of the **caudate nucleus** and **putamen;** the **globus pallidus;** the **substantia nigra;** and the **subthalamic nucleus.**
3. These structures are extensively interconnected with the **cerebral cortex** and **thalamus** to form two parallel but antagonistic circuits known as the **direct** and **indirect basal ganglia pathways.**
   a. Both pathways use **disinhibition** to produce their effects; that is, one population of inhibitory neurons inhibits a second population of inhibitory neurons.
   b. The **direct basal ganglia pathway** results in an **increased level of motor cortex excitement** and promotion of movement.
   c. The **indirect basal ganglia pathway** results in a **decreased level of motor cortex excitation.**
   d. Most neurons use **GABA** as their neurotransmitter, but connections between the **striatum** and **substantia nigra** use **dopamine.**
   e. Cholinergic neurons in the striatum release **acetylcholine,** which drives the indirect pathway, **decreasing cortical excitation.**

BASAL GANGLIA LESIONS (FIGURE 7–17) REVEAL FUNCTIONS OF THE BASAL GANGLIA

- ***Lesions of the striatum release inhibition*** *due to* ***degeneration of GABA neurons.***
  –These lesions occur in patients with ***Huntington chorea*** *and are characterized by* ***uncontrollable, quick, random movements*** *of individual limbs.*
  –Patients may also exhibit ***athetosis*** *(slow, wormlike, involuntary writhing movements) most noticeable in the fingers and hands.*
- ***Lesions of the globus pallidus*** *result in an inability to maintain postural support.*
- ***Lesions of the substantia nigra*** *are due to destruction of* ***dopaminergic neurons.***

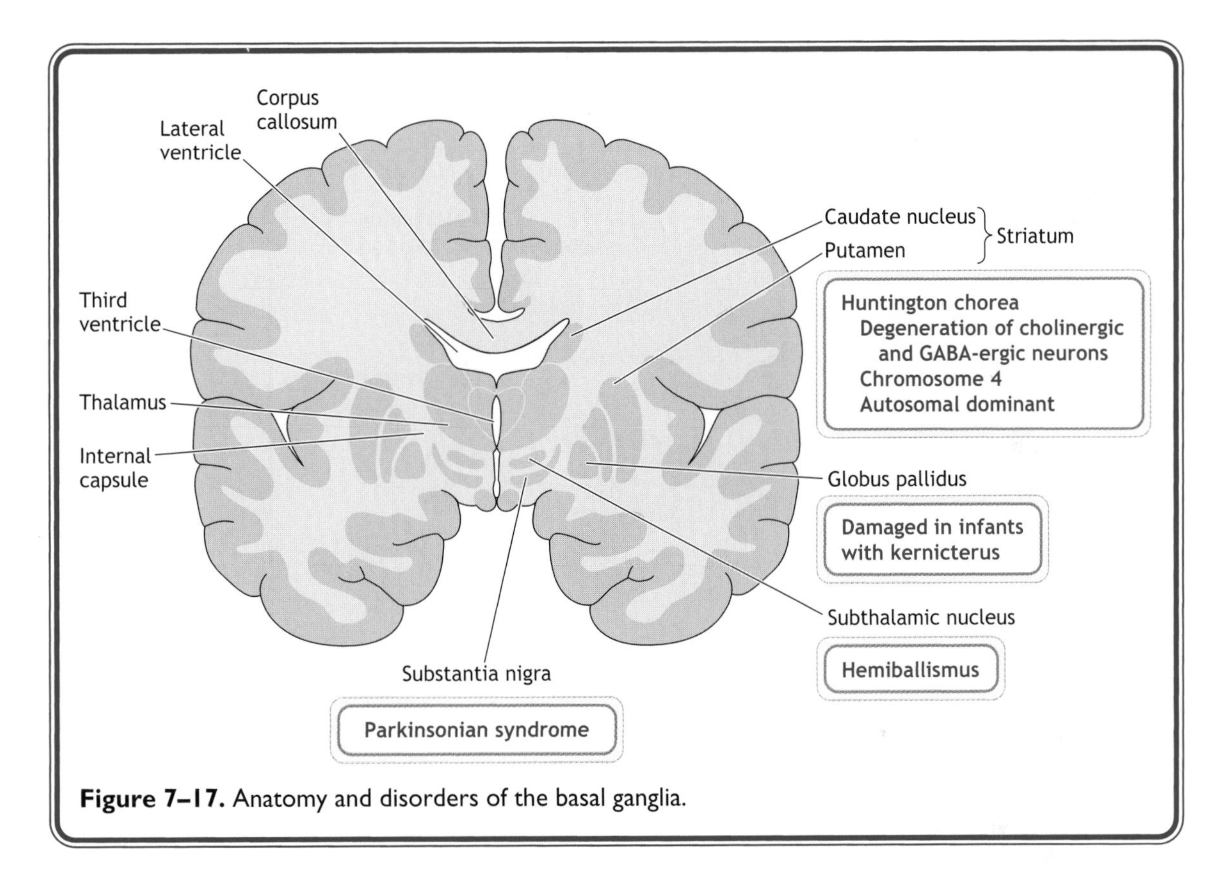

Figure 7–17. Anatomy and disorders of the basal ganglia.

–They occur in patients with ***Parkinson disease.***

–They are characterized by ***pill-rolling tremor of the fingers*** *at rest,* ***lead-pipe rigidity, and akinesia*** *(lack of voluntary movement).*

- ***Lesions of the subthalamic nucleus*** *are caused by hemorrhagic destruction of the* ***contralateral subthalamic nucleus.***

  –These lesions occur in hypertensive patients.

  –They are characterized by ***hemiballismus*** *(wild flinging movements of half of the body).*

IV. Language Function of the Cerebral Cortex (Figure 7–18)

A. The two cerebral hemispheres are not symmetrical morphologically or functionally.

B. Information is transferred between the two hemispheres of the cerebral cortex through the **corpus callosum.**

C. Most people (about 90%) are right-handed, which implies that the left hemisphere is more highly developed.

D. In most **right-handed people, speech and language** functions are also predominantly **organized in the left hemisphere.**

E. Most left-handed people show bilateral language functions.

F. Stimulation of some areas of the cortex elicits specific responses or sensations, whereas stimulation of other areas has no detectable effect. The latter are "silent" areas called the **association cortex.**

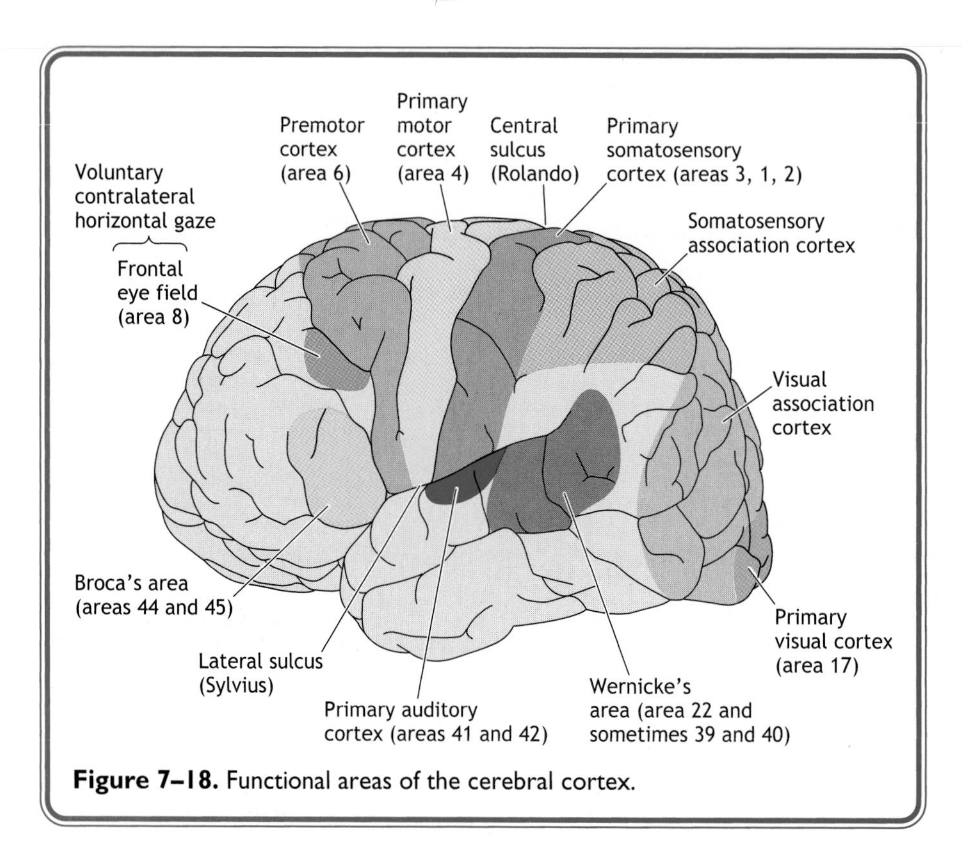

Figure 7–18. Functional areas of the cerebral cortex.

G. Language and speech are coordinated in specific areas of the **association cortex.**

1. **Broca's area** is a part of the **prefrontal** cortex or **frontal association cortex** in the dominant (left) hemisphere and is concerned with the motor aspects of speech (see Figure 7–18).
2. **Damage to Broca's area** produces a **motor** (**nonfluent**) **aphasia** or **expressive aphasia,** in which patients can understand language but have little ability to speak or write.
3. **Wernicke's area** is another important language area located in the posterior region of the temporal lobe next to the primary auditory cortex in the left hemisphere (see Figure 7–18).
4. **Damage to Wernicke's area** results in **receptive** (**sensory**) **aphasia,** in which patients have difficulty comprehending written or spoken language.
   a. Patients with **Wernicke's aphasia** often misuse words but are generally unaware of their deficit.

OTHER LESIONS AFFECTING LANGUAGE

- ***Gerstmann Syndrome***

  *–A lesion confined to the **angular gyrus** results in a **loss of the ability to comprehend written language (alexia)** and **to write (agraphia)**, but spoken language is understood.*

  *–**Fingeragnosia** (inability to recognize one's fingers) and right-left disorientation are present.*

- ***Conduction Aphasia***
  *–This disorder is due to a lesion in the **arcuate fasciculus.***
  –Patients are unable to repeat words or execute verbal commands but are otherwise verbally fluent.
  –Patients are frustrated by their inability to execute a verbal command they understand.
  *–This is an example of a **disconnect syndrome,** representing an inability to send information from one cortical area to another.*
  *–It may result from blockage of the **left middle cerebral artery branches.***
- ***Transcortical Apraxia***
  *–This disorder is due to a **lesion in the corpus callosum** caused by a **blockage of the anterior cerebral artery.***
  *–Patients cannot execute the command to move their left arm because a **corpus callosum lesion disconnects Wernicke's area** from the **right primary motor cortex.***
  *–Patients can still execute a command to move their right arm because **Wernicke's area can communicate with the left primary motor cortex without using the corpus callosum.***

V. The Blood-Brain Barrier and Cerebrospinal Fluid

A. Anatomy of the Blood-Brain Barrier

1. In most of the brain, capillary endothelial cells are connected by tight junctions **that prevent blood-borne substances from entering.**
2. **Astrocytes** have long processes with expanded vascular end-feet or **pedicels,** which attach to the walls of the capillaries to **maintain the blood-brain barrier.**

B. Functions of the Blood-Brain Barrier

1. The chemical integrity of the brain is protected by the blood-brain barrier so that a **constant environment is maintained** for neurons in the CNS.
2. The loss of CNS transmitters into the general circulation is prevented.
3. Water easily diffuses across the blood-brain barrier; non-ionized drugs cross more readily than ionized drugs.
   a. **Glucose,** the primary energy source of the brain, requires carrier-mediated transport; thus, the CSF has a lower glucose concentration than does blood.
   b. **Protein** and **cholesterol** are prevented from entering the CSF because of their large molecular size.

C. CSF Secretion and Distribution

1. Most **CSF** is **secreted by the choroid plexus.**
   a. The choroid plexus consists of glomerular tufts of capillaries covered by ependymal cells that project into the ventricles.
   b. The choroid plexus is located in parts of each lateral ventricle, the third ventricle, and the fourth ventricle.
2. CSF fills the **subarachnoid space** and **ventricles of the brain.**
3. CSF passes from the **lateral ventricles** through the interventricular **foramina of Monro** into the third ventricle. Then CSF flows through the **aqueduct of Sylvius** into the fourth ventricle.
4. CSF can leave the ventricles only through three openings in the fourth ventricle: **two lateral foramina of Luschka** and the median **foramen of Magendie.**

HYDROCEPHALUS (FIGURE 7–19)

Hydrocephalus is produced by an excess volume or by pressure of the CSF causing ventricular dilatation.

- ***Communicating Hydrocephalus***
  –This form of hydrocephalus is caused by excess secretion of CSF or by poor CSF circulation or absorption from the subarachnoid space.

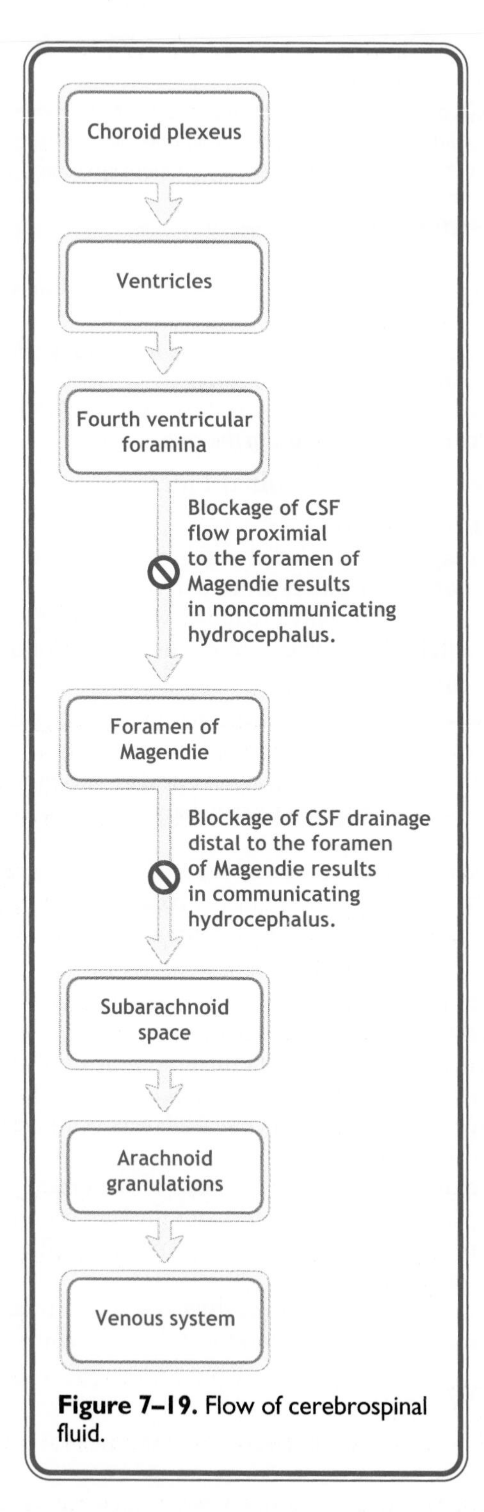

Figure 7–19. Flow of cerebrospinal fluid.

–It can be due to a tumor of the choroid plexus, a tumor of subarachnoid space blocking circulation, or meningitis inhibiting absorption.

- ***Noncommunicating Hydrocephalus***
  –This form of hydrocephalus is caused by obstruction of CSF flow inside the ventricular system.
  –CSF is prevented from leaving through the foramina of Luschka or Magendie; therefore, volume increases.
- ***Normal Pressure Hydrocephalus***
  –This form of hydrocephalus results from CSF not being absorbed by arachnoid villi and by ventricle enlargement pressing the cortex against the skull.
  –Patients exhibit confusion, ataxia, and urinary incontinence.

VI. Body Temperature Regulation

A. Body Temperature Values

1. **Normal body temperature** (from oral measurements) is 37°C (98.6°F).
2. An individual's temperature varies 0.5–0.7°C throughout the day.
3. Temperatures are lowest early in the morning and highest in the evening. This is called **circadian variation.**

B. Heat Production by the Body

1. The **specific dynamic action** (diet-induced thermogenesis) of **ingested food** appears to be due primarily to the digestion and assimilation of foodstuffs.
2. **Muscle activity** is a major factor in determining metabolic rate and heat production. If muscle activity is increased, heat production is increased, and vice versa.
3. The increase in metabolic rate and heat production by **catecholamines** is termed the **chemical thermogenic action.** When body temperature drops, the increased sympathetic discharge and increased release of epinephrine and norepinephrine from the adrenal medulla stimulate many processes, increasing heat production.
4. **Thyroid hormones** [triiodothyronine (T_3) and thyroxine (T_4)] **increase metabolic rate and heat production** by stimulating Na^+/K^+-ATPase activity. For example, in **hyperthyroidism,** temperature may be elevated 0.5°C, whereas in hypothyroidism, temperature may be depressed 0.5°C.
5. **Brown fat** is found in many young animals, including humans.
   a. Brown fat cells are richly innervated by sympathetic nerve fibers.
   b. Cold temperatures activate the sympathetic nervous system and activate β receptors in brown fat, thereby increasing metabolic rate and heat production.
   c. In human infants, brown fat may serve as a **physiologic electric blanket.**

C. Heat Loss

1. **Heat moves from the skin to the environment by radiation, conduction, convection, and evaporation**
   a. **Sixty percent of heat loss** is by **radiation** in the form of infrared heat waves when the ambient temperature increases.
2. Heat transfer by **conduction** occurs when the body touches solid material of different temperature.
3. Heat transfer by **convection** occurs when a fluid such as air or water carries the heat between the body and the environment.
   a. The conductive heat loss is proportional to the difference between the skin and ambient temperature.
4. Humans can dissipate nearly all of the heat produced during exercise by the **evaporation** of sweat from the skin surface.

a. Sweating is primarily a sympathetic cholinergic response in which postganglionic sympathetic fibers release acetylcholine to activate sweat glands.

D. Temperature Regulation Mechanisms

1. The **preoptic anterior hypothalamus and the skin surface contain temperature-sensitive cells that respond to changes in local temperature.**
   a. These cells increase their firing with increased temperature and decrease their firing with decreased temperature.
   b. They may defend against increased body temperature by stimulating sweating, vasodilation, and sympathetic outflow to sweat glands.
2. There are **skin receptors for cold and hot.**
   a. The ratio of cold-to-hot receptors is about 10:1.
   b. Interaction between cutaneous cold receptors and hypothalamic temperature-sensitive cells is thought to be responsible for the body's response to cold temperatures.
   c. Anterior hypothalamus temperature-sensitive cells send signals that inhibit the cold-response centers in the posterior hypothalamus and are especially important during exercise.
   d. With decreased temperatures, anterior hypothalamic cells decrease their firing and posterior hypothalamic inhibition is removed.
   e. Stimulation of **cold-response centers** causes **shivering, cutaneous vasoconstriction,** and **increased metabolism.**

E. Set Point

1. Body temperature variations initiate responses that bring the temperature back to normal, or to its **set point.**
2. This is similar to the thermostat setting on an air conditioner or heating unit.
   a. Thermal effectors include the cutaneous circulation, sweat glands, and skeletal muscles responsible for shivering.
   b. If the body core temperature is **below the set point,** the posterior hypothalamus **activates heat-generating mechanisms** (eg, shivering).
   c. If the body core temperature is **above the set point,** the posterior hypothalamus **stimulates heat loss mechanisms** (eg, vasodilation of cutaneous vessels).

F. Fever (Figure 7–20)

1. Fever is caused by circulating cytokines called **pyrogens** which are polypeptides produced by cells of the immune system.
2. **Endotoxin,** a cell-wall lipopolysaccharide of gram-negative bacteria, is a potent **exogenous pyrogen.**
3. Phagocytic leukocytes act on exogenous pyrogens to produce **endogenous pyrogens.**
4. Other endogenous pyrogens include **tumor necrosis factor-α** (TNF-α), **β and γ interferon** (β-IFN and γ-IFN), and **interleukin-6** (IL-6).
5. Endogenous pyrogens breach the blood-brain barrier located in the **organum vasculosum laminae terminalis** (**OVLT**) in the third ventricle.
6. Interleukin-1 beta (IL-1β) acts on cells in the OVLT to increase the release of **prostaglandin E_2**, which increases the set point.
7. Prostaglandin release explains the antipyretic (fever-reducing) property of aspirin.
   a. **Aspirin** is a **cyclooxygenase inhibitor** and blocks prostaglandin production, thereby decreasing the set point.
   b. **Steroids** reduce fever by blocking **arachidonic acid release** from brain phospholipids, thus preventing prostaglandin production.

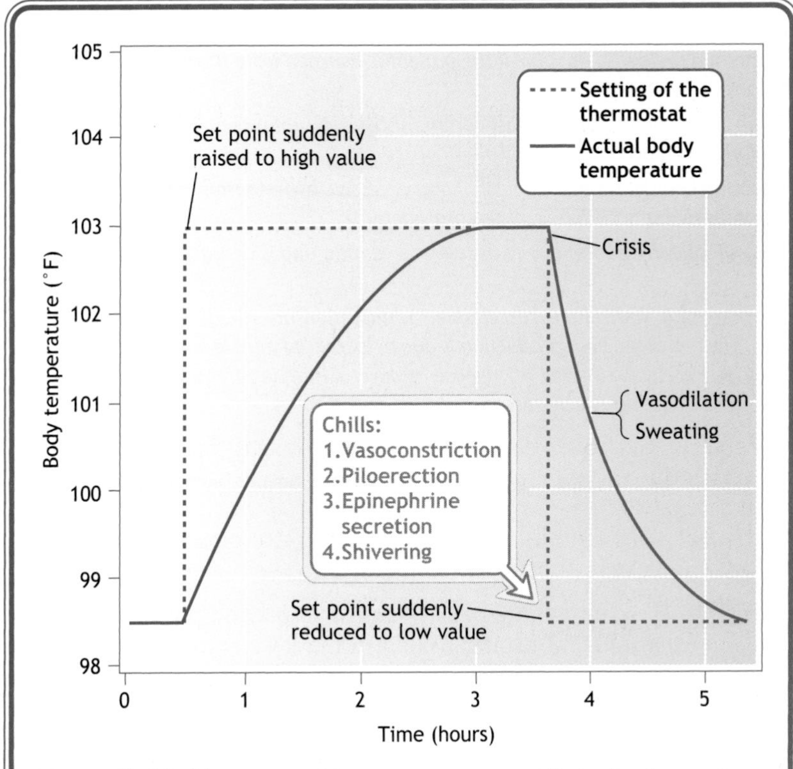

Figure 7–20. The onset of fever can occur rapidly in the form of a chill. The brain thermostat is raised suddenly, the person feels cold, and marked vasoconstriction and shivering occur. The combination of decreased heat loss and increased heat production increases body temperature up to a new set point. When the febrile agent is no longer active or present, increased vasodilation and sweating eventually return the set point to normal.

G. Cold-Induced Vasodilation

1. The initial response to a cold environmental temperature is usually **cutaneous vasoconstriction.**
2. As body surface areas cool, **vasodilation** can occur.
   a. This vasodilation may be a protective response that prevents freezing of the body surface (frostbite).
   b. The primary mechanism has been attributed to cold-induced paralysis of vascular smooth muscle.
   c. An example of cold-induced vasodilation in the human is the facial flush, or "rosy cheeks," of an individual on a cold day.

DISORDERS OF THERMOREGULATION

- ***Hypothermia*** *results when heat-generating mechanisms (eg, peripheral vasoconstriction and shivering) are unable to maintain body core temperature near the set point.*

–It is most commonly due to immersion in cold water.
–It is defined as a core temperature below 35°C.
–Recovery from extreme hypothermia (below) is possible if the patient is warmed from the inside out (eg, by warmed blood transfusions).

- ***Hyperthermia*** *(abnormally high core temperature) is most commonly due to prolonged exposure to heat and high humidity particularly when accompanied by physical activity.*
- ***Heat stroke*** *occurs when the body's heat loss mechanisms fail and excessive hyperthermia produces tissue damage (eg, vascular thrombosis, hemorrhage, and cerebral edema).*
- ***Heat exhaustion*** *may be the result of dehydration due to excessive sweating and is characterized by fatigue and dizziness.*
- ***Malignant hyperthermia*** *is observed in individuals susceptible to inhalation anesthetic agents or neuromuscular blocking agents. The increased heat production is due to increased muscle contraction and a hypermetabolic process triggered by excessive* Ca^{2+} *release in skeletal muscle. Treatment with dantrolene sodium reduces the* Ca^{2+} *release and decreases mortality.*

CLINICAL PROBLEMS

A woman brought into an emergency room is unresponsive and is displaying posturing (flexion of upper extremities and extension and plantar flexion in the lower extremities). Pupils are 4 mm in size and unreactive to light. No eye movements occur with head turning (oculocephalic maneuver) or with ice-water irrigation of the ear canals.

1. Which of the following is the most likely diagnosis?

A. Brain death

B. Hysteria-conversion coma

C. Brainstem hemorrhage

D. Drug ingestion

E. Bilateral internal carotid artery occlusion

A patient who complains of imbalance is found to walk with a wide-based gait and to sway forward and backward on standing. Balance cannot be maintained when standing with the feet together whether the eyes are open or closed. No limb ataxia or nystagmus can be elicited.

2. These findings are most consistent with a lesion in the

A. Vestibular apparatus

B. Midline vermis cerebellar zone

C. Pontocerebellar zone

D. Lateral cerebellar zone

E. Left frontal cortex

A 42-year-old man, who has had difficulty concentrating on his job lately, comes for medical attention because of irregular, jerky movements of his extremities and fingers. A sister and an uncle died in mental institutions, and his mother became demented in middle age.

3. Which of the following is the most likely diagnosis?
   A. Alcoholic cerebral degeneration
   B. Huntington chorea
   C. Wilson disease
   D. Hallervorden-Spatz disease
   E. Gilles de la Tourette disease

A 55-year-old woman is evaluated for weakness. Over the past few months she has noted slowly progressive weakness and cramping of her left leg. Lately she has also had some trouble swallowing foods. She is awake and alert. Findings on the neurologic examination are normal except for marked atrophy with fasciculations in the muscles of both legs, hyperactive reflexes in the upper and lower extremities, a diminished gag reflex, and a positive extensor plantar response.

4. Which of the following is the most likely diagnosis?
   A. Cervical spondylosis
   B. Guillain-Barré syndrome
   C. Lambert-Eaton syndrome
   D. Vitamin B_{12} deficiency
   E. Amyotrophic lateral sclerosis
5. Which of the following elements would be involved in the perception of pain due to an injurious stimulus?
   A. Spinocerebellar tract
   B. Spinothalamic tract
   C. Ventral horn of the spinal cord
   D. Red nucleus
   E. Nucleus ambiguus

A patient complains of hearing loss in the right ear. A 256-Hz tuning fork is positioned over the middle of the patient's forehead; the patient reports that he hears the tone in his right ear. He also notes better perception of a tone when the tuning fork is placed in contact with the right mastoid process than when it is placed outside his right ear.

6. Lesions in which of the following structures could account for these findings?
   A. Thalamus
   B. Central auditory pathways
   C. Cochlea
   D. External auditory canal
   E. Auditory cortex

A 9-year-old child is diagnosed with hyperopia.

7. In this child
   A. Rays of light from a point target at infinity converge in front of the retina in the unaccommodated eye

B. Accommodation may correct difficulties in distance vision
C. A concave lens will correct the refractive error
D. The primary cause is a malfunctioning ciliary muscle
E. Distant objects are fuzzy if the refractive error is 4 diopters or more

8. Bitemporal hemianopsia visual defects are associated with lesions of the
A. Pyramidal tract
B. Medial lemniscus
C. Occipital lobe
D. Optic nerve
E. Optic chiasm

ANSWERS

1. A is correct. Brain death is a clinical diagnosis of irreversible cessation of all cerebral and brainstem function. All brainstem reflexes are absent. Pupils are mid position and fixed. Vestibuloocular reflexes are absent. Muscle tone is flaccid with no facial movement and no motor response to noxious stimuli. Hysteria-conversion coma (choice B) is associated with decorticate posturing in response to noxious stimuli and is characterized by flexion of arms, wrists, and fingers and extension of the lower extremities. Brainstem hemorrhage (choice C) is associated with sudden loss of consciousness, quadriparesis, and pinpoint pupils. Drug ingestion (choice D), such as of cocaine or amphetamines, may be associated with subarachnoid hemorrhage, which can cause coma with signs of increased cranial pressure. Bilateral internal carotid artery occlusion (choice E) is associated with hemiparesis and aphasia.

2. B is correct. Midline vermis cerebellar zone lesions produce a wide-based gait and stance with posture instability. Vestibular apparatus lesions (choice A) are associated with vomiting, vertigo, and nystagmus away from the lesion side. Pontocerebellar zone lesions (choice C) are associated with ipsilateral facial paralysis and hearing loss. Lateral cerebellar zone lesions (choice D) impair limb movement ipsilateral to the lesion. Left frontal cortex lesions (choice E) produce contralateral sensory and facial weakness deficits.

3. B is correct. The patient's symptoms and age are consistent with Huntington chorea. Because the patient has jerky movement of his extremities and fingers, and relatives had dementia in middle age, a genetic cause is suggested. Alcoholic cerebral degeneration (choice A) is characterized by memory loss and ataxia. Wilson disease (choice C) is an autosomal recessive disease that causes a defect in copper metabolism resulting in copper overloading in the liver, cornea, and brain. Patients exhibit signs of parkinsonism, liver insufficiency, postural tremor, dystonia, and ataxia. Hallervorden-Spatz disease (choice D) is an autosomal recessive disorder due to a deficiency of cysteine dioxygenase, which leads to increased cysteine levels that promote free-radical formation, cell damage, and death. Symptoms occur before age 10 years. Gilles de la Tourette disease

(choice E) has an onset before age 21 years and is associated with multiple motor tics, one or more vocal tics, and a fluctuating course.

4. E is correct. Amyotrophic lateral sclerosis is a progressive degenerative disorder of the upper and lower motoneurons producing muscle weakness, spasticity, hyperreflexia (upper motoneurons), atrophy, fasciculations, and hyporeflexia (lower motor neurons). Cervical spondylosis (choice A) is an osteoarthritis involving the joints and discs of the cervical spine. Symptoms involve pain after motion and lifting. Patients with Guillain-Barré syndrome (choice B) report a tingling sensation in the arms and legs followed by rapidly progressive ascending symmetric muscle weakness. These patients have hyporeflexia of the extremities. Patients with Lambert-Eaton syndrome (choice C) exhibit weakness and fatigability of proximal muscles with depressed or absent tendon reflexes. Muscle strength may increase after exercise. Vitamin B_{12} deficiency (choice D) is characterized by generalized weakness and fatigability due to pernicious anemia.

5. B is correct. The spinothalamic tract conveys pain, temperature, and crude touch. The spinocerebellar tract (choice A) carries proprioception from the lower limbs to the cerebellum. The ventral horn of the spinal cord (choice C) contains principally α and γ motoneurons whose motor axons innervate skeletal muscle. The red nucleus (choice D) is a globular mass located in the ventral portion of the tegmentum of the midbrain and acts as a relay center for many of the efferent cerebellar tracts. The nucleus ambiguus (choice E) is a cigar-shaped nucleus in the medulla that innervates the voluntary muscles of the pharynx via CNs IX and X and of the larynx via CN X.

6. D is correct. The patient is suffering from conduction deafness because the vibrations of the tuning fork are conducted better by bone (mastoid) than by air (next to the ear). A tuning fork placed next to the forehead will result in sound being localized in the affected ear. Unilateral deafness is not associated with central lesions such as in the thalamus (choice A), central auditory pathways (choice B), cochlea (choice C), or auditory cortex (choice E).

7. B is correct. Hyperopia, or farsightedness, is caused when the eyeball is shorter than normal and the parallel rays of light are brought to focus behind the retina. Choice A is incorrect because in hyperopia, rays of light are brought to focus behind the retina, not in front of it. Choice C is incorrect because a biconvex lens, not a concave lens, will correct hyperopia by adding to the refractive power of the lens of the eye. Ciliary muscle malfunction (choice D) is not associated with hyperopia. Choice E is incorrect because in hyperopia, distant objects are clear but near objects are fuzzy.

8. E is correct. Lesions affecting the chiasm disrupt crossing fibers from the nasal halves of both retinas, causing bitemporal hemianopsia. Choice A is incorrect because the pyramidal tract is made up of axons from the posterior frontal and anterior parietal cortical areas that terminate in the spinal cord. Choice B is incorrect because the medial lemniscus is composed of fibers from the gracile and cuneate nuclei of the medulla that ascend to the thalamus, carrying information on pressure, limb position, vibration, direction of movement, recognition of texture, and two-point discrimination. Choice C is incorrect because the occipital lobe is involved primarily with visual perception and involuntary smooth pursuit eye movements. Choice D is incorrect because lesions of the optic nerve impair vision from the ipsilateral eye but do not cause bitemporal hemianopsia.

INDEX

Note: Page numbers followed by *t* and *f* indicate tables and figures, respectively.

Notes